ADVANCES IN CONDENSED MATTER AND MATERIALS RESEARCH

VOLUME 3

ADVANCES IN CONDENSED MATTER AND MATERIALS RESEARCH
VOLUME 3

FRANCOIS GERARD
EDITOR

Nova Science Publishers, Inc.
New York

Senior Editors: Susan Boriotti and Donna Dennis
Coordinating Editor: Tatiana Shohov
Office Manager: Annette Hellinger
Graphics: Wanda Serrano
Book Production: Matthew Kozlowski, Jonathan Rose and Jennifer Vogt
Circulation: Cathy DeGregory and Ave Maria Gonzalez
Communications and Acquisitions: Serge P. Shohov

Library of Congress Cataloging-in-Publication Data
Available upon request.

Includes bibliographical references and index.
ISBN 1-56072-147-8

227 Main Street, Suite 100
Huntington, New York 11743
Tele. 631-424-6682 Fax 631-425-5933
e-mail: Novascience@earthlink.net
Web Site: http://www.nexusworld.com/nova

Printed in the United States of America

CONTENTS

PROSPECTS FOR NEW ALUMINIUM ALLOYS BASED ON MULTIPHASE EUTECTICS

N.A. Belov and V.S. Zolotorevskii

The method of creation of cast aluminium alloys of the new generation – a natural composites based on the multiphase eutectics containing transition and rare-earth metals (Fe, Ni, Ce, Cr and others) has been theoretically substantiated. The essence of method is that during the transition from two-component eutectics to multiphase ones, one can obtain the fine-differential structure with inclusion volume fraction of second phases more than 20 vol.%, i.e. this structure is near that has been realised in quickly-cooled alloys, giving possibility to obtain the foundry of the same running abilities. It has been developed and tested the quick searching method based on investigating the results of the only alloy, that allows sharply to increase the experimental investigations efficiency.

INTRODUCTION

An aluminium alloys thanking to unique combination of the operational, technological and economical merits are the most spread constructional materials. In last time the requirements of diminishing the weight of products are toughened not only in the aircraft building but also in other branches, particular in motor-car construction, stimulating to use the existent aluminium alloys and to develop the new ones. It should be noted that alloys based on other light metal (titanium and magnesium) will hardly compete with aluminium among the product of repetition work, that is connected with their higher cost and technological difficulties of production. Based on this tendency, a lot of investigations directed on the creation of new alloys composite materials on the base of aluminium of the improved properties are carried out at present time.

Analysis of researches of the last decades [1-7] shows that essential improvement of required characteristics of the commercial aluminium alloys of both the cast alloys and the deformed alloys by means of small modifications of the contents and the processing conditions (heat treatment and shapig) will not most likely succeed. Use of special technologies (supercrystallisation, mechanical alloying and others) allows essentially to increase different properties, however, because of high cost, their practical use is very limited. The essential changes (i.e. the decrease of prime cost to the level allowing to expect the repetition work) in the nearest time is not expected for a while. In this connection, a development of the fundamental new aluminium alloys, firstly the cast alloys, intended for traditional technology, can be a prospective direction. In particular, the way based on the using of multiphase eutectic alloys on the base of aluminium, having fine-differential structure (thickness of dendritic branches is less of 1 μm) with the inclusion volume fraction of second phases up to 20-30 vol.%, i.e. being near this, that is realised in quickly–cooled alloys [4], that allows expecting the possibility of obtaining the foundry of the same level of service properties, in particular, the characteristics of heat-resistance and corrosion resistance with keeping a high level of low-temperature mechanical properties.

THE IDEA OF A PROPOSED APPROACH

It has been shown in work [8-15] that in two-phase aluminium alloys the transition from hypoeutectic alloys to eutectic ones is followed not only by the increase of inclusion volume fraction of second phase but also by dispersion of eutectic colonies. As the result of latter effect, many second phases contained low-soluble elements (Fe, Ni, Ce and others) in aluminium solid solution, become able to the fragmentating and the spheroidizing at the homogenizing annealing. From this it follows that eutectic alloy has optimal structure and, respectively, can provide the best complex of different properties:

- hypoeutectic alloys contain a less quantity of second phase and respectively, has a lower level of those properties, which increase with increasing the inclusion volume fraction of this phase;
- because of existing the rough first crystals, the hypereutectic alloys have low mechanical properties (first of all, the indices of plasticity and breaking strength).

Unfortunately, in the foundry even the best two-phase eutectic structures do not allow to reach the level of strength and heat-resistance of quickly-cooled steels. The reason of this is that two-phase eutectics which are realisable in foundry have either second phase inclusion volume fraction limitation (not more than 10-15 vol.%), or eutectic colony dispersion limitation (the thickness of branches of second phase is more than 1 μm at the hardening rate of 10 K/s). One of the way of elimination of these defects, i.e. the increase of Q_V up to 20-30 vol.% and decrease of dendritic branches thickness on the exponent (i.e. to 0.1-0.5 μm), is the transition from two-phase to multiphase eutectics. Its essence is as the following.

In triple systems based on Al (Fig.1) during the transition from two-component eutectics (e_1 and e_2) to three-component eutectic (E), the total concentration of second and third components (B1 and B2) is increasing, as a rule. The consequence of that is the increase of total inclusion volume fraction (ΣQ_V) of eutectic phases (EP) β_1 and β_2 in alloy E. Besides, three-component eutectics in comparison to two-component ones, are superfine. One can assume that transition to four- and more component eutectics must lead to a higher value of ΣQ_V and a finer structure of eutectic, i.e. the transition to multiphase eutectics may have the same influence on structure as the increase of the cooling rate during crystallisation with all the positive sequences followed from this:

- fine-differential structure of eutectics phases and their high inclusion volume fraction ;
- possibility of influence on dimension and morphology of eutectic phases particles by means of homogenizing annealing in wide range of temperatures.

Thus, the use of multiphase eutectics allows to create the natural composites, structure of which consists of plastic aluminium matrix and solid inclusion of second phases (or fine-differential crystals, which are continuos within one eutectic colony). At that, the matrix features can be controlled by means of alloying the aluminium solid solution and its following precipitation hardening. In order to obtain a desired second phases of specified properties, one should choose a proper base systems which have multiphase eutectics. A possible types of such structures are shown in Fig.2.

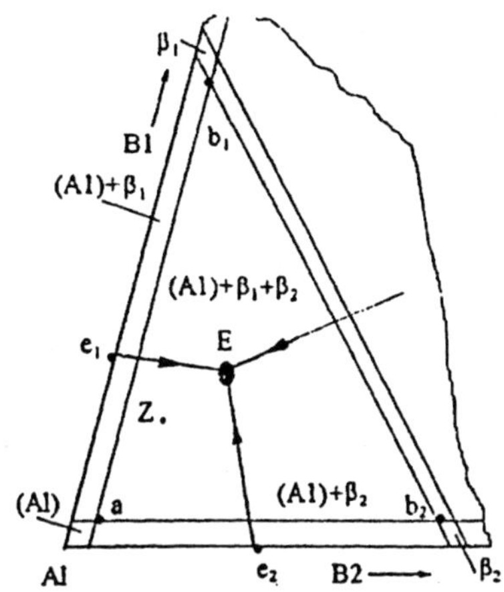

Figure 1 An aluminum angle scheme of triple system Al-B1-B2: (Al) is solid solution on the base of aluminum; B1 and B2 are the alloying elements; β_1 and β_2 are a second phases based on binary compounds of type of (Al,B1) and (Al,B2)

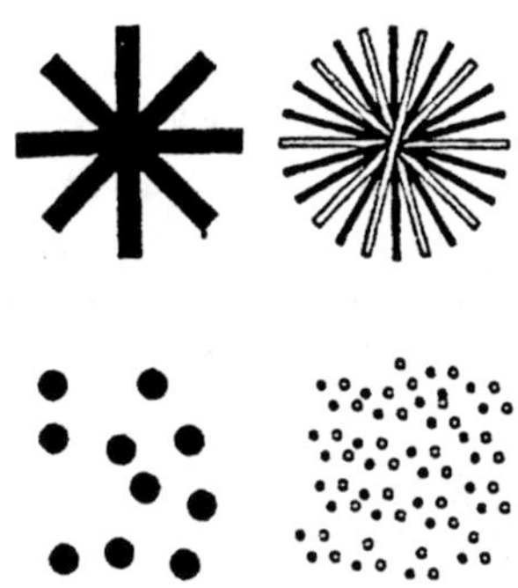

Figure 2 Scheme of the eutectic alloys microstructures; a, c – two-phase eutectics (Al)+β_i; b, d – three-phase eutectics (Al)+β_1+β_2; a, b – casting state; c, d – heat-treated state

As far as data about equilibrium diagrams (ED) of quadruple systems (not to mention the ED with great number of components) based on the aluminium are not numerous, the confirmation of this assumption requires a significant work to build them. The efficiency of this work can significantly be increased by carrying out the preliminarily analysis of a known ED and by revealing the prospective systems and the concentration regions in which one can expect multiphase eutectics with desired structure. Having limited at first stage by building the three- and four-component ED and, respectively, by a known phases which create the two-component eutectic with aluminium solid solution, one can experimentally reveal eutectic alloys and evaluate their structure when there are comparatively not many alloys. If the received results are not satisfactory ones, than the building of new EDs (rather single, the most important their elements), in which the odds of the searching out of new eutectics with desired structure are large enough, will be a next stage. As the approach which makes easy the searching out, one can assume that all phases in multi-component systems are created on the base of compounds with the number of components more than 4, that is followed from the analysis of literature [17-19]. It is obvious, that because of the limitations of economical character, some elements of periodic table, which are interested from scientific point of view (all noble and radioactive metals, germanium, most of rare-earth metals), should be excluded from consideration. The final result of this work should be a contents of a prospect basic multiphase compositions on the base of which one can develop the casting and deformable alloys of the new generation.

Among two-component ED of eutectic type [16,18] the alloys Al-Si, Al-Ni, Al-Ce and Al-Fe(metastable) are of the greatest interest from point of view of the inclusion volume fraction, morphology and EP properties. Concerning the cobalt and manganese, they are of a lesser interest, because of small value of Q_V of phases Co_2Al_9 and $MnAl_6$. A lanthanum, neodymium and other rare-earth metals which interact with aluminium analogously to cerium, but are rarely used for alloying the aluminium alloys, are of a small interest as well. The systems with main alloying elements (Mg, Cu, Zn) do not approach because of their too high concentration in the eutectic alloys, and the system with lithium, calcium and other alkali-earth metals because of the technological difficulties of fusion.

The two-phase eutectics can be created not only in double systems, but also in quasi-binary profiles of the multi-component ED. As there are very few information about the studied quadruple ED [16-19], and the prospective quasi-binary profiles were not revealed at them (besides of system Al-Si-Fe0-Be [12]), experimental confirmation of the said above is only touched to triple systems. Among the latter the most interesting is the following: Al-Mg-Si, eutectic (Al)-Mg_2Si; Al-Fe-Ni, eutectic (Al)-$FeNiAl_9$; Al-Fe-Ce, eutectic (Al)-Fe_2CeAl_{10}, Al-Ce-Si, eutectic (Al)-$CeSi_2Al_2$; Al-Fe-Si eutectic (Al)-Fe_2SiAl_8, nonequilibrium variant. The basis characteristics of all the considered above two-phase eutectics, which are of practical interest, are presented in Table 1. It should be noted that because of absence of sufficiency evidences in literature, the contents of a some eutectic alloys are approximate and are basically relayed on results of experiments carried out earlier by authors. The values of inclusion volume and mass fraction of EP (Q_V and Q_M) were obtained by calculation in respect with contents and phases density.

The comparison characteristics of basic parameters (Q_V and morphology EP, temperature of solidus, mechanical properties and others) of eutectics alloys, given in Table 1, shows that the best candidate for the formulated task is the alloy Al-6%Ni, for which at crystallisation rate of 1-20 K/s, the average diameter of dendritic branches of phase $NiAl_3$ is 0.5-1 μm, that ensures almost complete passing of spheroidization processes already at 450^0C [8-11]. At the same time the other compositions possessing individual advantages (economy, for example, at the presence of iron) can be considered as well.

Two-phase eutectic alloys, given in Table 1, were taken as the initial alloys for solving the task of the searching out the multiphase alloys. This part of work included not only the analysis of perspective ED and obtained earlier experimental data, but also the special experimental investigations, in order to have the possibility of comparing the different alloys under equal conditions.

Table 1 Main characteristics of three-phase eutectics on the base on aluminum

#	Eutectic phase	System	Contents and eutectic crystallization temperature	Q_M, mass.%	Q_V, vol.%
1	(Si)	Al-Si	12% Si, 577°C	10.5	10.9
2	Mg_2Si	Al-Si-Mg	8.5% Mg, 4.5% Si, 595°C	16.5	11.5
3	$NiAl_3$	Al-Ni	6% Ni, 640°C	14.3	9.7
4	$CeAl_4$	Al-Ce	12% Ce, 640°C	19.6	13.1
5	$FeAl_6$	Al-Fe (metastab.)	2 %Fe, 655°C	7.7	6.5
6	$FeNiAl_9$	Al-Fe-Ni	1.75% Fe, 1.75% Ni, 650°C	10.2	8.1
7	Fe_2CeAl_{10}	Al-Fe-Ce	2% Fe, 2.5% Ce, 650°C	9.1	5.7
8	Fe_2SiAl_8	Al-Fe-Si (unequal)	3% Fe, 2% Si, 630°C	9.4	7.1
9	(Ce,Si,Al)	Al-Ce-Si	8% Ce, 2% Si, 640°C	14	7
10	$Fe_4Si_2Be_3Al_{16}$	Al-Fe-Si-Be	3% Fe, 1% Si, 0.4% Be, 640°C	9	8

EXPERIMENTAL TECHNIQUE

The investigated alloys have been produced in electric furnace in graphite-chamotte crucibles from pure components (aluminium A99, magnesium Mg90, semiconductor silicon) and ligatures Al-20%Ni, Al-10%Fe, Al-10%Mn, Al-10%Cr, Al-20%Ce, Al-5%Be and some others, the contents of which was thoroughly analysed. The investigations were carried out on the ingots of dimension of 10×20×180 mm and 15×30×180 mm, obtained by means of the founding in graphite molds (V_{cool}=10 K/s). The ingots were thermally treated in electric muffle furnace with the accuracy of temperature maintenance of 3 K. For obtaining the slowly-cooled alloys the lumps of ingots of weight of 10-15 g were remelted in alundum crucibles and cooled together with muffle furnace ($V_{cool}=10^{-2}$ K/s). The direct and thermal analysis was carried out on samples of mass of 5-10 g according to the standard techniques with the accuracy of determination of critical temperatures of 1-2 K.

The structure of alloys was investigated on light and electron scanning microscopes (JSM-35 CF). At latter the reflection electron regime was mainly used in order to obtain maximum phase contrast. On microscope JSM-35 CF with the help of 4-crystal wave spectrometer the micro-X-ray analysis of the phase content was also carried out. Metallographic sections were made by the mechanical polishing for the preliminary reviewing and by electrolytic for the detailed analysis in scanning electron microscope, revealing the structures of dimension up to 0.05 μm.

MULTIPHASE EUTECTICS

From analysis of triple ED [17-19] it follows that in the frames of the proposed approach (Fig.1) one should choose a triple-phase eutectic alloys of a good studied systems Al-Si-Mg and Al-Si-Ni. It is obvious that there should be a lot more of the potential multiphase eutectic alloys with a searched structure. That is confirmed by the results of the study and the particular building of multi-component ED, carried out under the leadership of authors during last years (Table 2) [8, 12-14]. In particular, in system Al-Si-Mg-Ni [13] not only quasi-triple section of eutectic Al-$NiAl_3$-Mg_2Si has been revealed, but also quarter eutectic (Al)-(Si)-$NiAl_3$-Mg_2Si, in which the total inclusion volume fraction of EP does not exceed 20 vol.%, and their morphology is favourable enough. It was searched out also the triphase eutectics (Al)-EP1-EP2, where EP1 are phases (Si) or Mg_2Si, and EP2 is Fe-contained phases: $FeNiAl_9$, $Fe_9Si_2Be_3Al_{16}$ and Fe_2SiAl_8. It is obvious that on the base of latter eutectics, one can develop the most economy alloying alloys.

However, for developing the high-temperature alloys (this direction turns to be the most prospective), all the considered eutectics do not approach, mainly because of low solidus (lower 575^{0}C) and presence of easily diffusing Mg and Si at working temperatures (300-400^{0}C) [4,5,9].

One can avoid these lacks by using only transition and rare-earth metals (TM and RM, respectively) which create eutectic ED with aluminum. In particular, the eutectic (Al)-$NiAl_3$-$CeAl_4$, reached in triple system Al-Ce-Ni [14], is of high-dispersion structure building and total inclusion volume fraction of EP is more than 20 vol.% (Fig.3), that indicates about the appropriateness of the considered early approach. One can assume with large probability that similar triple eutectics can be also searched out in other systems Al-Ni-REM. This assumption is based on similar character of interaction of different REM with Al [16-18] that become apparent in isomorphic correspondence of many two- and three-component compounds of type (Al, REM) and (Al, REM, TM). As cerium (along with lanthanum) is the cheapest and the most accessible REM, the study of the rest of ED of considered type on a given stage is baseless. The use of misch metal consisting basically of Ce and La is probably the optimal variant.

In this connection the choice of the potential alloying elements which, in principle, can create high-temperature multiphase eutectics, is essentially constricted: Ni, Ce, Fe, Co and Mn. Unfortunately, in other triple systems with participation of chosen eutectic elements similar to (Al)-$NiAL_3$-$CeAl_4$ was not searched out, therefore, the question of particular building of quarter and more complicated ED on the base of aluminum is sharply arising. Since the fundamental study of even one quarter system is a complicated and expensive investigation and, respectively, can not be justified within frames of searching work, the development of technique of fast determination of basis characteristics of multi-component system becomes actual one.

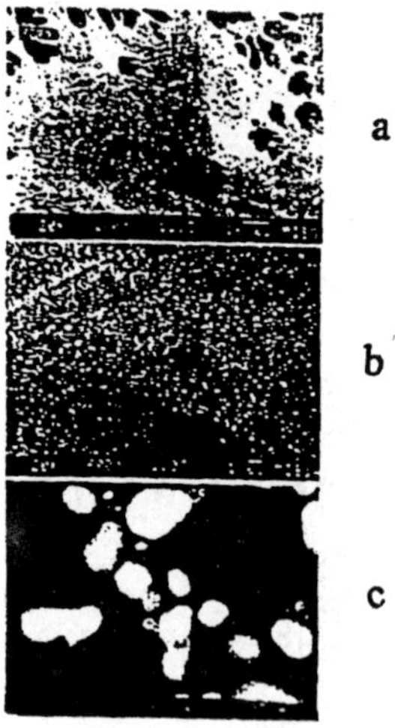

Figure 3 Structure of alloy Al-12%Ce-5%Ni, REM, reflected electrons: a, b – the casting state, fine-differential eutectic (Al)+$CeAl_4$+$NiAl_3$, c – heat-treatment, 600^{0}C, 12 hours, the compact inclusions of the phases $CeAl_4$ (white) and $NiAl_3$ (gray) against a background (Al), magnification: a–×1000; b–×6000; c–×10000

Table 2 Fundamental Characteristics of Three-Phase Eutectics on the Basis of Aluminium

#	Phase Effect	System	Composition and Temperature for Crystallization of Eutectics	Q_M, mass.%	Q_V, vol.%
1	(Al)-(Si)-Mg_2Si	Al-Si-Mg	Al-13% Si-5% Mg, 555°C	16.5	19.2
2	(Al)-(Si)-$NiAl_3$	Al-Si-Ni	Al-11%Si-4.5%Ni, 565°C	20.5	18.0
3	(Al)-Mg_2Si-$NiAl_3$	Al-Si-Mg-Ni	Al-4%Si-7%Mg-3%Ni, 588°C quasi-triple section	16.5	18.2
4	(Al)-(Si)-Mg_2Si-$NiAl_3$	Al-Si-Mg-Ni	Al-12%Si-4%Mg-2%Ni, 548°C	19.2	20.1
5	(Al)-(Si)-$FeNiAl_9$	Al-Si-Fe-Ni	Al-5%Si-1.5%Ni-1.5%Fe, ≈575°C	13.0	11.4
6	(Al)-(Si)-$Fe_4Si_2Be_3Al_{16}$	Al-Si-Fe-Be	Al-6%Si-2%Fe-0.3%Be, ≈575°C	10.4	10.0
7	(Al)- Mg_2Si-$FeNiAl_9$	Al-Ni-Fe-Mg-Si	Al-1%Ni-1%Fe-7%Mg-4%Si, ≈590°C, quasi-triple section	14.1	16.9
8	(Al)- Mg_2Si-Fe_2SiAl_8	Al-Si-Fe-Mg	Al-3%Si-2.5%Fe-2%Mg, ≈575°C (monovariant) unequally weighted variant	10.7	10.3
9	(Al)- Mg_2Si- $Fe_4Si_2Be_3Al_{16}$	Al-Fe-Si-Mg-Be	Al-2%Fe-3.5%Si-4%Mg-0.25%Be, ≈590°C, quasi-triple section	11.3	12.2
10	(Al)- $NiAl_3$-$CeAl_4$	Al-Ni-Ce	Al-12%Ce-5%Ni, 627°C	33.0	24.9

TECHNIQUE OF QUICK OBTAINING OF MULTI-COMPONENT EQUILIBRIUM DIAGRAM

As far as within the frames of a put problem, the searching out of multiphase eutectics with favourable structure is one of the key stage, the ED analysis can be limited by solving the following questions:

- is there the multiphase eutectic of the disperse structure in a given system?
- what is its content and what the melting (crystallization) temperature is?
- phases of which contents are in the composition of this eutectic?

In principle, one can answer all these questions by investigating only alloy using the behaviors of multicomponent ED of the invariant eutectic.

Let us choose the tree-component hypoeutectic alloy Z (Fig.1) as the simplest example. In this alloy a some quantity of the triple eutectic (Al)+β_1+β_2 obviously must be of the morphology which is typical for merely eutectic alloy E that allow to answer the first question. The contents of this eutectic can be easily determined by micro-X-ray analysis using scanning regime, and the melting temperature can be determined by thermoanalysis technique. Further it should be performed the annealing of the alloy Z near eutectic temperature with aim of spheroidization and coalescence of particles of EP, that allow from one hand to evaluate the formability of the studied eutectic, and from other hand to obtain the particles which are large enough for the microanalysis of their contents. The results of the latter allow to determine the peaks of eutectic triangle a-b1-b2 and, respectively, to build the aluminum angle of triple ED (Fig1). The proposed method has been tested on the system Al-Ce-Ni with positive result, i.e. the contents of eutectic alloy appears to be close to this, that was early determined by means of the traditional obtaining of this ED [14].

In order to obtain an additional information about considered ED, it is advisable to analyze the structure of slowly-cooled alloy Z, that is caused by two reasons:

- a rougher structure of eutectic makes easy to identify EP;
- it is appeared the possibility of evaluating the degree of equilibrium crystallization deviation.

The latter is connected with that under real conditions of the founding (particularly in chill mould), the nonequilibrium crystallization takes place, which in a number of cases can strongly influence not only on the position of eutectic point E (usually shifting it into the side of the increase of B1 and B2), but also on the phase content (this is related to a more complicated systems in comparison to Fig.1). For example, in system Al-Si-Fe the fine-differential eutectic (Al)+Fe_2SiAl_8 is formed only under equilibrium conditions at the cooling rate not less than 10 K/s, when the $FeAl_3$ and $FeSiAl_5$ is suppressed, which must be crystallized according to the equilibrium ED Al-Si-Fe. Respectively, a lot 3- and 4-phase eutectics with participation of Fe_2SiAl_8 can also be realized only under conditions of strong equilibrium concentration deviation. A more detailed influence of content and cooling rate with reference to triple and quadruple systems is considered in work [15].

Provided that all the parameters of triple ED studied by a proposed method are satisfying, the preparation of alloy of merely eutectic contents must be the next stage in order to determine the basic properties and, respectively, to make a final decision about its prospect for a detailed research. The said above is entirely related to a quadruple system (determination of peaks of 4-phase tetrahedron) and a more complicated systems, which can not be graphically imaged (simplexes in n-dimensional space), but can be analyzed by calculation techniques.

It should be noted that EP can be not only binary compounds, as shown in Fig.1, but also 3 and 4-componwent phases, that is seen in Table 1-2. On one hand, that complicates the analysis of ED, and on other hand essentially increases the chances of searching out the eutectics with the desired parameters. For example, in aluminum angle of the system Al-B1-B2 with two triple compounds γ_1 and γ_2, in principle, there can be triphase eutectics (Al)+β_1+γ_1; (Al)+γ_1+γ_2 and (Al)+γ_2+β_2 and 2 quasi-binary sections (Al)+γ_1 and (Al)+γ_2. In 4-component systems the number of possible eutectic combinations is much larger, especially at presence of 4-component compounds.

The analysis of a known triple and quadruple ED shows that the systems of merely eutectic type are met not very often. Most systems, as rule, have peritectic reactions that decrease the total number of possible eutectic combinations and complicates the analysis of a system, especially for the conditions of nonequilibrium crystallization. However, taking into account the possibility of existence of 3 and 4-component compounds, the number of potential alloying elements can significantly increase, first of all due to TM creating peritectic ED with aluminum (Cr, V, Mo and others), as well as small additions of Si and Be.

The silicon, which was culled on the first stage of the triple system analysis, can be useful in multi-component ED provided that silicon is engaged in the contents of the complicated phases of the type (Al,Si,TM) or (Al,TM1,TM2), i.e. is absence in (Al) and is not crystallized in form of (Si). It should be separately marked out the beryllium which is engaged into some industrial cast aluminum alloys for neutralizing the deleterious effect of impurity of iron [1-2,12], that is connected with creation of multi-component phases of type (Al,Fe,Be) or (Al,Fe,Si,Be). From analysis of ED, one can conclude that in the presence of other TM the Berilium can lead to the creation of phases of type (Al,TM,Be), (Al,TM,Be,Si) and (Al, TM1, TM2, Be) thereby increasing much more a chances to find the eutectic alloys of desirable structure. However, because of a high toxicity of the oxide, the practical use of beryllium is very limited and, respectively, its consideration as potential alloying element must be limited as well.

Summarizing a said above, one can propose the algorithm of carrying out the research works, including both fundamental and applied researches, that is shown in Fig.4. The works which are necessarily required for finding the eutectic alloys are marked with bold, and the desirable works which can give a useful information for the further ED construction are marked with fine.

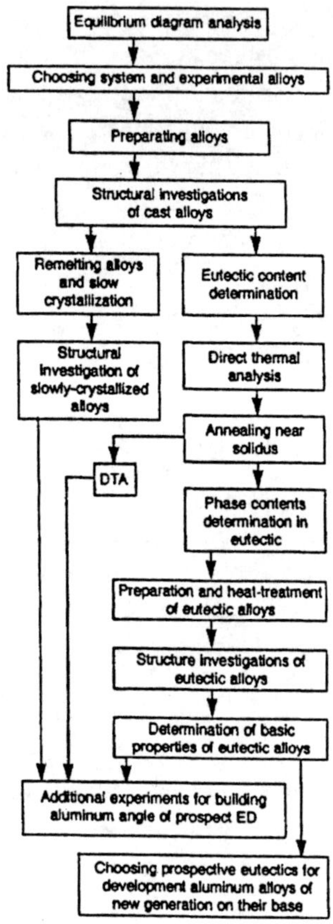

Figure 4 The algorithm of searching the multiphase eutectics for developing the aluminum alloys of new generation on their base.

The preliminary experiments carried out on alloys with addition of Fe, Ni, Ce, Si, Mn, Mo, Co, Cr and Be have demonstrated the efficiency of a given technique, because of the short time of the working and the low material facilities expenses, one can solve the question of the expediency of further researches. It allows, on one hand to reach quickly practical results and, on other hand to concentrate fundamental investigations in the direction of prospective developments in order to create a scientific basis for further progress in designing the aluminum alloys of new generation.

The said is illustrated by microstructures of one of the alloy X* with multiphase eutectic (Fig.5). Despite of the structure of a given alloy is very far the ideal one, the presence of fine-differential eutectic Y* (Fig.5a) makes the investigated system prospective, that is additionally confirmed by the structure of eutectic Y after the spheroidizing (Fig 5b). Thus, the preparation of only one alloy Y (containing more than 1% of Fe) and determination of its properties can lead to the determination of contents of basic composition, based on which, one can develop a heatproof cast aluminum alloys of new generation with operational characteristics on the level of quick-cooled eutectic alloys, in particular, of 01489 and CV78 on the base of systems Al-Fe and Al-Fe-Ce, respectively.

* The alloy X is hypoeutectic one, the total contents of the alloying elements is about of 5 mass%; in the eutectic alloy Y the total contents of elements is about of 10 mass.%.

Conclusion

1. The method of creation of natural compositions based on multiphase eutectic aluminum alloys during the use of traditional foundry technology has been theoretically substantiated. The essence of method is that during the transition from two-component eutectics to multiphase ones, one can obtain fine-differential structure (dimension of particles is less of 1 μm) with inclusion volume fraction of a second phases up to 30 vol.%.
2. The technique of the quick searching of multiphase eutectics according to the results of investigation of a given alloy has been developed and experimentally tested, that allows to increase sharply efficiency of experimental investigations.
3. The prospect of multiphase eutectics as basic structural component of alloys of systems on the base of aluminum with transition metals (Fe, Ni, Co, Mn, Cr and others) and rare-earth elements (first of all Ce and La), as well as small additions of silicon and beryllium, has been demonstrated.

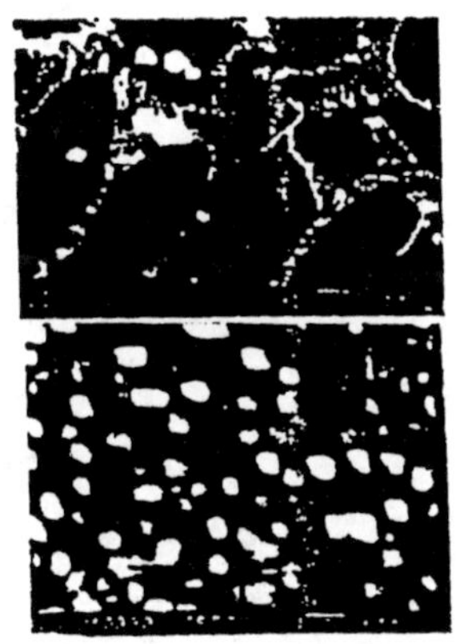

a

b

Figure 5 Structure of alloy X of system Al-Fe-Ni-Ce-Cr-Mn-Si with multiphase eutectic Y, REM, reflected electrons; a – cast state, a small amount of fine-differential eutectic Y; b – heat-treatment, 600°C, 12 hours, compact inclusions at least of 3 phases of eutectic Y against the background (Al). Magnification: a - a–×1000; b - a–×10000.

References

1. Aluminiy. Svoystva i fizicheskoye metalovedenie: Sprav., J. E. Hetch edition. Moscow, Metalurgia, 1989, 324
2. Promyshlennyye aluminiyevyye splavy. Sprav, izd. Alieva S.T., Altman, M.B., Ambatsumyan S.M. et al,. Moscow : Metalurgiya, 1984, 528
3. Proc.5th Int.Conf.on al-alloys and Their Physical and Mechanical Properties (ICaa5), 1-5.07.96 Grenoble, France, Materials Science Forum, v.1-4.
4. Dobatkin V.I., Elagin V.I Bystrozakrystalizovannyye aluminiyevyye splavy. Moscow, VILS, 1995, 341 - in Russain
5. Kolobnev I.F. Zharoprochnost liteynyh alyumi-nievyh splavov. M.: Metallurgiya, 1973,. 320 s.
6. Elagin V.I. Tehnologiya legkih splavov, 1998, no.1, s.7-23
7. Stroganov G.b. Vysokoprochnye liteynye alyumi-nievye splavy. M. Metallurgiya, 1985.
8. Belov N.A. Izv.vuzov. Tsvetnaya metatturgiya, 1995, no.1, s.48-57.

9. Belov N.A. Journal of Advanced Materials, 1994, v. 1 (4), p.321-329.
10. Belov N.A., ZolotorevskiiV.S., Luzgin D.V. Journal of Advanced materials, 1996, v.3(3), p.228-238.
11. Belov N.A. Metallovedenie i termicheskaya obrabotka metallov, 1993, no.10, s. 19-22.
12. Belov N.A., Kurdyumova T.A. Izv.AN SSSR. Metally, 1989,no.2,s.210-215
13. Belov N.A. Metally, 1993, no.5, s.133-138.
14. Belov N.A, Naumova E.S. Metally, 1996, no.6, s.146-152.
15. Belov N.A. Metally, 1995 , no.1, s.44-51.
16. Diagrammy sostoyaniya dvoynyh metallicheskih sistem. (Spravochnik), t. 1,. N.P.Lyakishev, O.A.Bannyh, L.L.Rohlin i dr. M.: Mashinostroenie, 1996, 992 s.
17. Diagrammy sostoyaniya sistem na osnove alyuminiya i magniya. (Spravochnik) M-E.Drits, N.R.Bochvar, E.S.Kadaner idr.M.:Nauka,1977,228s.
18. Mondolfo L.F. Struktura i svoystva alyuminievyh splavov. Per. s angl. M.: Metallurgiya, 1979, 640 s
19. Diagrammy sostoyaniya metalicheskih sistem. Vyp.1-XXXVIII, M.:VINITI, 1959-1994.

ABOUT THE AUTHORS

Belov Nikolai Alexandrovich *– Moscow institute of steels and alloys (Technological University), Doctor of technical science, professor. An expert in the area of physical metallurgy of aluminum alloys*

Zolotorevskii Vadim Semenovich *- Moscow institute of steels and alloys (Technological University), Doctor of technical science, professor. An expert in the are of physical metallurgy of aluminum alloys and composite materials.*

FORMING INTERPARTICLE CONTACTS AND REDISTRIBUTING PARTICLES UNDER LASER SINTERING OF METAL-POLYMER POWDERS MIXTURE

N.K.Tolochko, V.B.Mikhailov, S.E.Mozzharov, N.V.Sobolenko, I.A.Yadroytsev and M.I.Zhemchuzhnyi

The kinetics of forming the interparticle contacts and the process of redistributing the particles during laser sintering of metal-polymer powders mixture has been theoretically studied. The correlation in the character of development of both processes has been demonstrated.

INTRODUCTION

The technology of the selective laser sintering is comparatively new direction in powder metallurgy [1-3]. At present time, two-component powders, in which one of the component plays the role of the fusible binding material, are widely used in this technology. Thermoplastic polymers are usually used as the binding materials. Therefore the study of the sintering behavior of such powders under laser emission is of great interest. The aim of present work is the experimental investigation of kinetic of forming the interparticle contacts and redistributing particles under laser sintering of metal-polymer powders mixture.

The raise of such an experiments has a great importance for developing not only the technology of selection laser sintering but also as a entire technology of powder metallurgy. However, up to now the performance of these experiments under the conditions of conventional sintering with using the furnace equipment was connected with certain technical difficulties. These difficulties can be practically excluded provided that the sintering is carried out under laser emission. Use of laser as the heating source allows to accomplish such an experiments setup, under which the very suitable conditions for observing the particles behavior during their sintering are ensured.

FORMING INTERPARTICLE CONTACTS

Nickel alloy particles (10-20Cr; 2.8-3.4B; 4-.5Si; 0.6-1C; Ni is the restt) (M) and polymer based on epoxypolyester compounds (P) were the subject of inquiry. The particles were of spherical shape of diameter of 300-450 μm. They were obtained by means of the laser remelting of a smaller particles. In every experiment three particles of approximately equal diameter were undergone to sintering under effect of continuous Nd laser: YAG of type of LTN-103 (λ=1.06 μm, of power of 250 W) of a fixed density of flux radiation Q. The particles in every three were placed abreast due to scheme M-P-M on plain transparent quartz glass substrate. During experiments the main parameters characterizing the forming of interparticles contacts were determined, such as distances between centers of an extreme particles, L; diameter of middle particle, D; the average diameter of the contacting chords for two extreme particles, d ($d=(d_1+d_2)/2$). The particles sintering in every experiment was accomplished within frames of single continuous cycle. The behavior of sintering particles was recorded with the help of optical microscope connected with video and photo apparatus.

The particles bonding on the difference stages of the sintering is schematically shown in Fig. 1. A typical behavior of the parameters of forming interparticles contacts is presented in Fig.2.

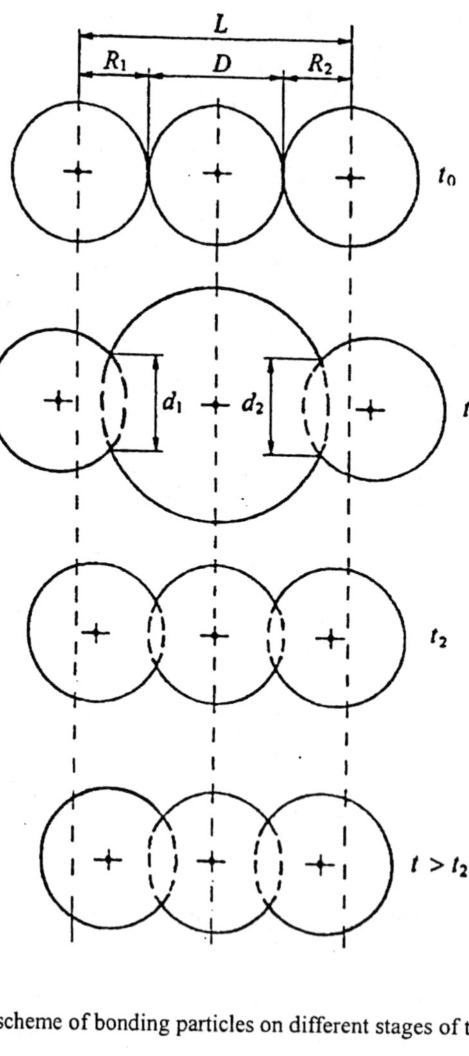

Fig.1 The scheme of bonding particles on different stages of the sintering

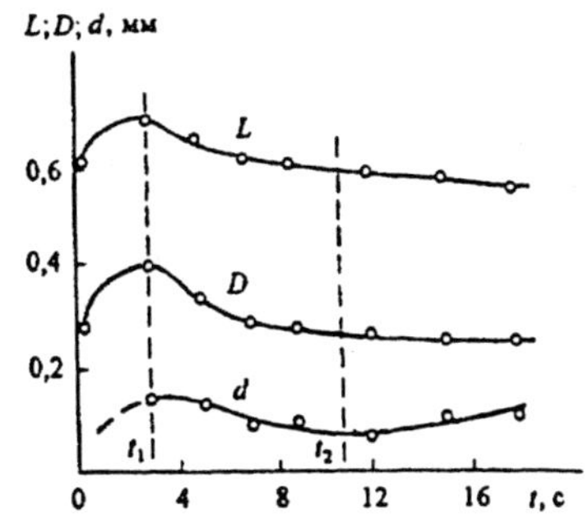

Fig.2 The change of distance L between centers of extreme particles, diameter D of middle particle and diameter of the contacting chords d during forming interparticle contacts. Q=95 W/cm^2.

This process includes the following stages, which are typical for liquid-phase sintering of two-component powders: 1) melting (transition into plastic state) of low-melting polymer particle; 2) forming the liquid-phase bonding between extreme solid particles of metal; 3) shrinking of solid particles with each other; 4) the polymer melt consolidation and the forming of rigid polymer constraint between metallic solid particles. At the same time this process has some a peculiarities.

The melting of particles of polymer is followed with its "swelling" (D_2 increases). This can be connected with gassing. At the same time on this stage L and d increase. At certain moment t_1, parameters D, L and d attain the peak values ($D_m \approx 0.4$ mm, $L_m \approx 0.68$ mm and $d_m \approx 0.15$ mm). Here $L_m < D_m + 2R \approx 0.71$ mm, where $R \approx 0.155$ mm is average radius of extreme particles($R = (R_1 + R_2)/2$). That is evidence of that together with mutual removal of extreme particles due to increasing of the polymer particles dimensions during its melting, the mutual shrinking occurs as well. Further (at $t > t_1$) D, L and d decreases. This decrease occurs comparatively quickly, and after is slowed down. At certain moment t_2 parameter D possesses one's initial value ($D_0 \approx 0.29$ mm), L becomes some lower of one's initial value L_0 ($L_0 = D_0 + 2R \approx 0.6$ mm), and d attains certain minimum (≈ 0.1 mm). Further (at $t > t_2$)D remains unchanged, while L goes on decreasing, and d, vice versa, starts increasing.

Such behavior of parameters D, L and d can be explained in the following way. The polymer melt is badly wetting the metal particles. As the consequence, the shrinking of extreme particles due to capillary phenomena is slight. In particular, that is confirmed by comparatively small value d. Therefore polymer particle, increasing in dimensions during melting, "pushes aside" the extreme particles. As the result of, L and d increase together with D. Vice versa, when the polymer drop decreases in dimensions it has not time to "drag up" quickly enough the metal particles to itself. As the result of, L and d are decreasing together with D. After D possessing a initial value and not changing further, the shrinking process goes on. As the result of, d begins again to increase, and L to decrease as before.

Redistributing Particles

The experimental technique was in many respects analogous to the technique, described above. The same particles of nickel alloy of predominantly spherical shape (dimension of 100-160 μm) and polymer particles of arbitrary shape (dimension of ≤500 μm) were the subject of inquiry. The sintering particles were mixed and placed predominantly as the monolayer. The laying of particles within the monolayer was certainly of randomicity. The different places of monolayer could have a different extent of density and regularity of arrangement. The spot of laser beam was usually covering whole area of monolayer. The density of radiation energy flow Q was comparatively uniformly distributed over the surface of monolayer. In different experiments the value of Q was varied (Q=80; 95 W/cm^2). Besides, the volume contents of metal-polymer ratio M:P was changing (M:P=1:1; 2:1; 3:1).

The typical series of images of redistributing particles during the sintering is shown in Fig.3 and 4. Here a white shapeless spots correspond to the pores. They should not be confused with white dots that are the flashing on the metal particles.

As seen from the figures, redistributing particles is characterized by their very complicated displacements followed with the change of porous microstructure of monolayer. If the particles in the layer at initial state are distributed comparatively uniformly, then during the sintering this uniformity usually maintains. If the initial layer has evidently nonuniform distribution particles density then during the sintering this nonuniformity usually increases. In this case the appearance of zonal isolating particles is possible.

a b c d

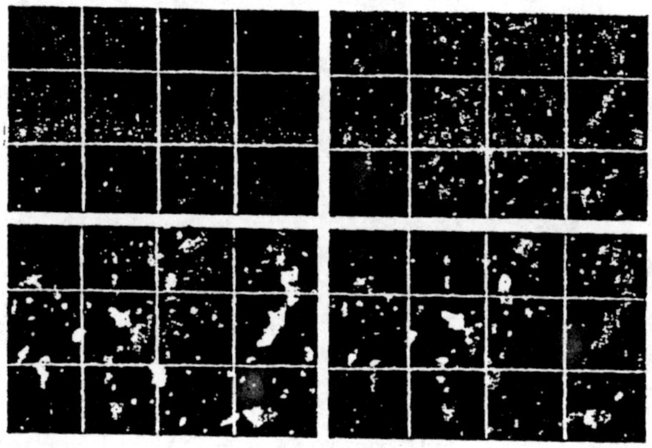

Fig.3 The pictures of redistributing particles during sintering. M:P=2:1; Q=80 W/cm^2. Sintering time, s: a –0; b – 4; b – 8; d – 10.

In general, analysis and quantitative estimation of redistributing particles process are quite a complex. This task was solved by us in the following way. The field of monolayer of particles for every picture of their redistribution was divided into individual sections (cells). The cells were the squares of the side being approximately equal to 6D, where D is average diameter of metal particles. The cells were placed by consecutive raw, entirely surrounding the area of all the monolayer or its the most characteristic part. In particular, every of pictures of redistributing particles shown in Fig. 3 and 4, contains 12 cells. Such a difragmentation of monolayer has allowed to control the peculiarity of redistributing particles on the individual monolayer sections, to compare each to other, to reveal their general and distinguishing features, to ascertain their correlation. As it was mentioned above, the redistributing particles is characterized firstly by changing porous structure of monolayer. Therefore the porosity S (the ratio of pores area within the cell to total cell area) was chosen as basic parameter of the process of redistributing particles. Thus, the porosity was determined not for whole monolayer field, but only for its individual cells.

a b c d

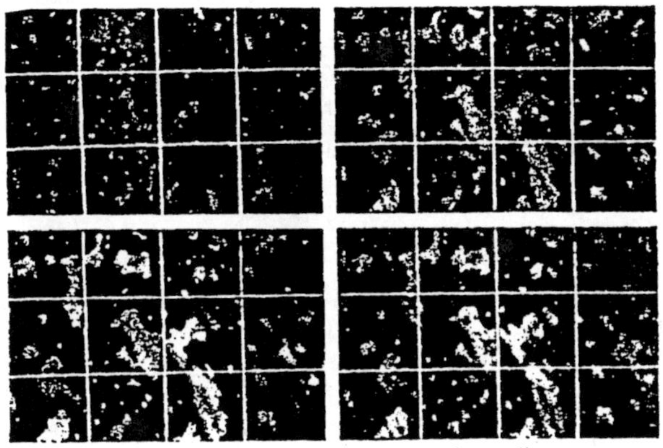

Fig.4 The pictures of redistributing particles during sintering. M:P=2:1; Q=95 W/cm^2. Sintering time, s: a –0; b – 2; b – 4; d – 5.

The typical behavior of the porous structures changes during redistributing particles is presented in Fig.5 and 6 (as diagrams of dependence S(t) for various monolayer cells.). These figures describes the redistributing processes which are shown as the pictures series in Fig.3 and 4, respectively.

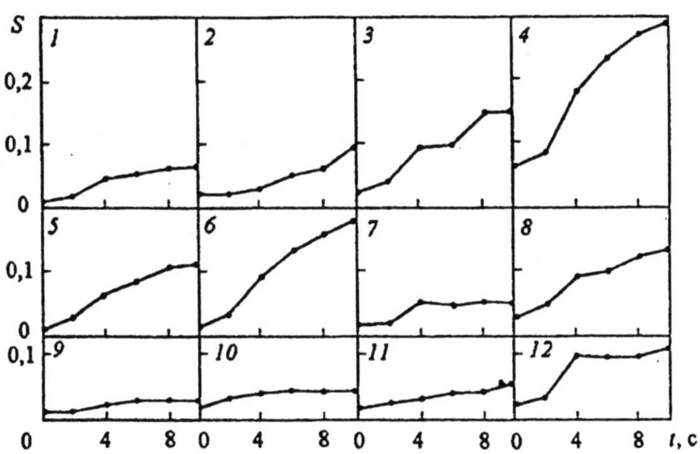

Fig. 5 Changing porosity S during redistribution. One correspond to redistributing particles, shown in Fig.3.

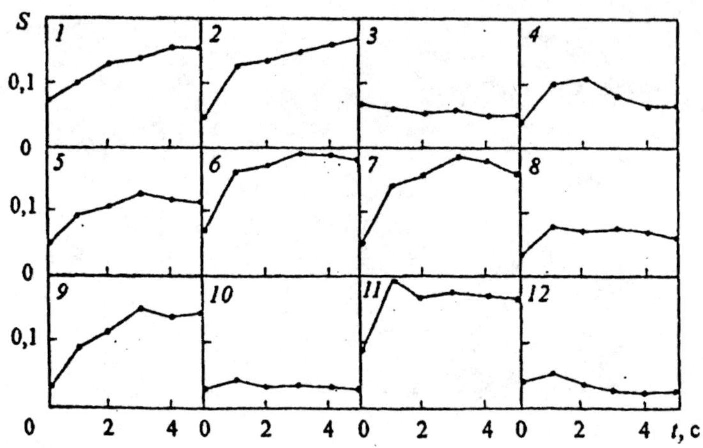

Fig. 6 Changing porosity S during redistribution. One correspond to redistributing particles, shown in Fig.4.

On the base of analysis of received results, one can emphasize the following the most generals behaviors of redistributing particles.

1. Within almost the whole monlayer, S increases at the beginning, reaches the peak value S_m and then decreases (Fig.5). On different places of monolayer S changes non-uniformly. Usually S_m is greater on the places, which have a higher initial porosity S_0. The rate of mentioned changes S increases with increasing the heating, i.e. density of energy flux of radiation Q (compare Fig.5 and 6).
2. The increase of S is rarely followed with the appearance of new pores. As rule, the growth of initially existed pores and individual pores join into larger pores is observed. Usually the pores are joining each with other, resulting the borders extension. The initial pores position is slightly changing.
3. Despite of the increase of S, the density of local particles position increases as well. As rule, the particles are closing in with each other within the individual monolayer sections, which firstly were surrounded with pores. On such a places the zonal isolating particles occurs. In general the particles are characterized with comparatively low mobility.

The pointed behaviors were characteristic for all the experiments independently of ratio M:P.

The considered peculiarities of redistributing particles are closely connected with peculiarity of forming interparticles contacts. It is good seen from comparison of diagrams dependencies L(t) (Fig.2) and S(t) (Fig.5 and 6). At the beginning of process D is increasing. It leads to the increasing of L and, as consequence, of S. Then D is decreasing. Respectively L and, as consequence, S, is decreasing as well.

CONCLUSION

An experimental results show that there is the correlation between the character of processes of forming interparticle contacts and character of redistributing particles processes during laser sintering. It is confirmed, in particular, by experimental investigations of kinetics of laser sintering of the metal-polymer powder mixtures. Thus, during the forming interparticle contacts the polymer particles diameter playing the role of the bonding material, and as consequence the distances between bonding metal particles are increasing at the beginning, and then are decreasing. During redistributing particles the porosity of powder layer changes analogously. As rule, at the beginning it increases as well, and then decreases. The effect of increasing the stated parameters at the initial stage of sintering is caused by swelling polymer resulting from the heating. The redistributing particles is followed by their zonal isolation, the degree of which depends on initial distribution of particles in layer. The received results can be used in practice of selective laser sintering.

REFERENCES

1. Tolochko N.K. Fizika i himiya obrabotki materialov, 1995, no.1,s.94-101.
2. Tolochko N.K., Sobolenko N.V., Mozzharov S.E. i dr. Vestnik mashinostroeniya, 1995, no.4, s.22-25.
3. Tolochko N.K., Mozzharov S.E., Sobolenko N.V. i dr. Poroshkovaya metallurgiya, 1995, no.3/4, s.32-37.

ABOUT THE AUTHORS

Nikolai Konstatinovich Tolochko– *Institute of Technical Acoustics of NAS of Belorussia (Vitebsk city), Doctor of Physics and Mathematics, head of laboratory. An expert in arae of physics and technology of sintering processes, crystallization and laser treatment of materials.*

Victor Borisovich Mikhailov– *Institute of Technical Acoustics of NAS of Belorussia (Vitebsk city),. An expert in the area of physics and technology of sintering processes.*

Sergei Evgenievich Mozzharov - *Institute of Technical Acoustics of NAS of Belorussia (Vitebsk city),. An expert in the area of physics and technology of sintering and crystallization processes.*

Nikolai Vasilievich Sobolenko - *Institute of Technical Acoustics of NAS of Belorussia (Vitebsk city), Candidate of Mathematics and Physics,. An expert in the area of physics and technology of sintering processes.*

Igor Anatolievich Yadroytsev– *Institute of Technical Acoustics of NAS of Belorussia (Vitebsk city),. An expert in the area of physics and technology of sintering processes and materials laser treatment processes.*

Mikhail Igorevich Zhemchuzhnyi– *Institute of Technical Acoustics of NAS of Belorussia (Vitebsk city), Candidate of Mathematics and Physics. An expert in the area of physics and technology of laser and plasma treatment of materials.*

Reflecting Coating for Global Spacecraft on the basis of Zirconium Dioxide Modified by Silicon Dioxide

V.M. Vladimirov and M.M. Mikhailov

The technique of production of modified pigment on the base of ZrO_2, consisting of the mixing and sintering of ZrO_2 and SiO_2 powders has been proposed. The pigment has been obtained, which is resistant to electromagnetic radiation of spectral distribution close to the sun due to the formation of optical contacts and decrease of light scattering near boundaries of grains.

The reflecting coatings on the base of pigment ZrO_2, having significant forbidden gaps (E_g=5.28 eV at 300 K [1]) are prospective for systems of passive temperature control of spacecraft, since due to developed technology of the production, the integral absorption of solar radiation is not more than 0.15, that is much less than for coatings on the base of such known pigments as ZnO, TiO_2 and Zr_2TiO_4.

The resistance of light-reflecting coatings to the ultra-violet radiation depends on series of factors: chemical compositions of pigment and the binding agent, the pigment granulometric composition, radiation power and radiation dose, the presence of either modifiers. Polycrystalline ZrO_2, characterized as light-resistant and atmosphere resistant pigment [2], has shown oneself as a perspective pigment for solar reflectors.

For stabilizing ZrO_2, a number of modifiers have been proposed: SrO [3], $SrSiO_3$ [4]. $SrNO_3$ and SiO_2 [5]. The main idea of these works was the possibility of forming the ZrO_2 particles of shielding coating, either due to reaction

$$ZrO_2 + SrSSiO_3 \xrightarrow{973-1273K} SrZrO_3SiO_2 \qquad (1)$$

or due to molecular layering [6]

$$SiCl_4+(n+2)H_2O \quad SiO_2{\times}nH_2O+4HCl \qquad (2)$$

Hence, according to authors [4], during photolysis, the surface and volume vacancies are behaving differently, number of which can be controlled by temperature of preliminary sintering. It is also known [5], that growth of the shielding film SiO_2 of more than one layer does not cause the significant increase of optical properties stability. This fact was not discussed in work [5], but it follows from formula (2) that crystalline hydrate is formed on the surface of ZrO_2 crystallites. Decomposition of $SiO_2{\cdot}nH_2O$ occurs during the sintering, and multilayer film is partially or completely destroyed, at that the grain face is uncovered and the modifier plays role of the ordinary microaddition.

It is known that boundary of pigment-binding agent creates the complex microrelief. The light passing through this microrelief causes not only refraction and reflection but also diffuse scattering (S) [6, 7]. For turoid mediums instead of Frenel's formula, the following expression is used

$$\frac{K}{S}=\frac{(1-R_{\infty})}{2R_{\infty}} \qquad (3)$$

where R_{∞} is reflectance of infinitely thick layer, K is absorption factor.

Value K is practically independent of pigment dispersity [8], and light-scattering S is determined by optical homogeneity and by number of optical contacts of pigment particles, i.e. by particles size distribution function. The size effect was also observed for zirconium dioxide. It has been shown that grain size influences on optical degradation process. Therefore the variation by granulometric composition allows achieving the optical homogeneity and obtaining the optical contacts with SiO_2 on the surface ZrO_2 resulting from preliminary sintering.

In a given work, the regimes of laying the SiO_2 on ZrO_2 powders were developed with the purpose of obtaining the pigment and reflecting coatings resistant to solar radiation and the charged particles of space. It have been used ZrO_2 powders of specification OSC 7-4 and SiO_2 powders of specification OSC of quantity 2-4%, which were intermingled in china bowl on roller mill during 6 hours.

X-ray analysis was carried out to apparatus DRON-3 (Cu K-radiation, nickel filter), survey angle was $15<2\theta<75^0$. These samples were prepared by laying the powder on object-plates covered with gluing agent.

Infrared absorption spectra were recorded on UR-10 apparatus in range of 400-4000 cm^{-1}. Samples were prepared by mixing it with KBr and by means of pressing the tablets.

The choosing of the sintering regimes was carried out using the data of DTA, which were obtained from "Paulic & Erdey" device. The mass of button was 150-200 mg, the heating rate was 10 degree/min.

The exposure to the solar light was carried out to "Spectr-1" plant in vacuum of not less than 10^{-5} Pa. Radiation dose corresponds to 170 e.s.d. (1 e.s.d. is equivalent solar day, being 0.14 W/cm^2. Radiation monitoring was performed with respect to ultraviolet region of radiation spectrum of DKSR-3000 lamp imitating radiation spectrum of the Sun. The principal scheme of plant and irradiation technique are described in [10].

According to spectra of diffusion reflection the integrated reflection figure a_s was calculated

$$a_s=1-\frac{1}{\lambda_1-\lambda_2}\int_{\lambda_1}^{\lambda_2}\rho_{\lambda d}\,d\lambda \qquad (4)$$

where ρ_λ is spectrum of diffusion reflection in range λ_1-λ_2.

An absorption, induced by radiation, was determined from difference of values of coefficients a_s for an irradiated and initial sample

$$\Delta a_s=a_{s1}-a_{s0} \qquad (5)$$

Powders were the heterogeneous mixture consisting of granules, grains and lamellas (Fig.1). Granulometric composition of mixture was studied to scanning electron microscope

EMV-10B using replica technique. Grain size was within 0.5-1.5 μm, granule size was within 5-250 μm, lamella length was up to 100-150 μm.

After sintering at temperature 300-1000^{0}C (Fig.2), particular interaction of pigment particles and modifier particles, as well as increase of number of contacts between grains ZrO_2 and SiO_2 took place. The choice of sintering regimes was carried out with respect to curve DTA (Fig.3), on which two peaks are clearly seen: the low-temperature peak at 200^{0}C is connected with water removal, high-temperature peak at 920^{0}C demonstrates the particular sinetring of two phases.

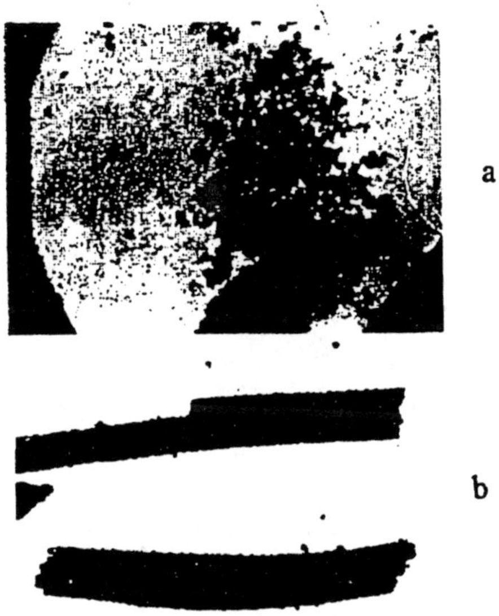

Fig.1 Granulometric composition of ZrO_2: a - solder pads (×35000); b –grains and granules (×12000)

The results of the measurements of integral absorption (Fig.4) show that with increasing the content of SiO_2 the value Δa_s decreases. Hence the resistance to the sunlight is some greater, than in [5]. Such a improvement of stability can be caused, on one hand by using zirconium dioxide of specification OST 7-4 (OST 9-2 powder was used in [5]) and by different granulometric composition, and on other hand by increasing the quantity of optical contacts of grain due to high SiO_2 concentration.

During the exposure, a several types of defects are created, which behave oneself in different ways (they are accumulated, migrated, annihilated and the light energy dissipation occurs at the same time). That is probably why the increase of modifier quantity can be not only reason of increasing the resistance of the coating to the sunlight.

For clearing up such changes, the analysis of reflection spectra difference was carried out, and exhibited in Fig.5. On the curve 1, which corresponds to the free additions ZrO_2, the absorption bands at 280, 390-420 nm are clearly exhibited. The curve for sample with 9% of SiO_2 is of the same absorption character, but is of smaller band intensity.

The bands in region of waves length <300 nm are near edge of fundamental absorption of ZrO_2 and are connected with appearance of Zr^{4+}, Zr^{3+}, Zr^{2+}, Zr resulting from particular dissociation of ZrO_2 molecule. The bands in region of 400 and 600 nm are caused by occurrence of F- and F'-colouring centers and by metallization of surface [4,5,9,11], thus, with accumulating the defects, band of 400 nm increases and satiates.

Fig.2 The effect of the sintering on aggregates states at temperature 900^{0}C during 6 hours: a - solder pads (×22000); b – aggregates (×35000)

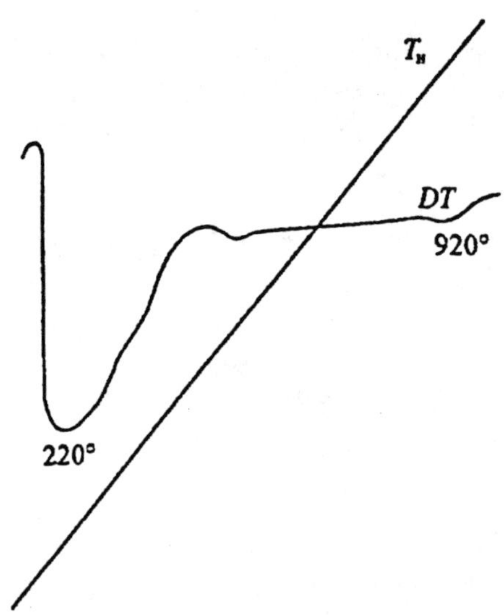

Fig.3 Curves of DTA of ZrO_2+2%SiO_2

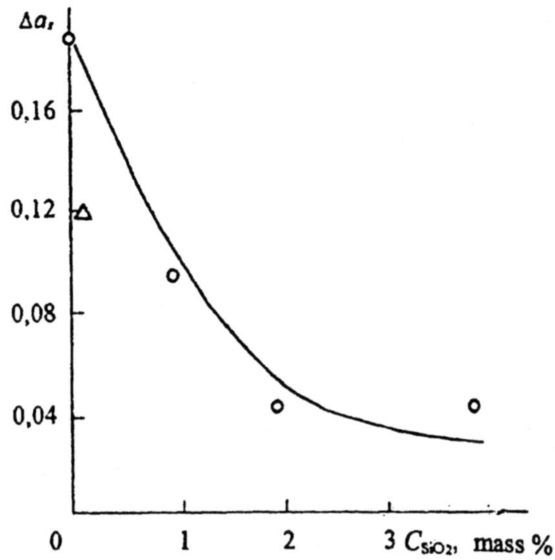

Fig.4 Dependence of Δa_s vs. concentration of SiO_2 in ZrO_2 exposed to solar light of dose of 170 e.s.d.

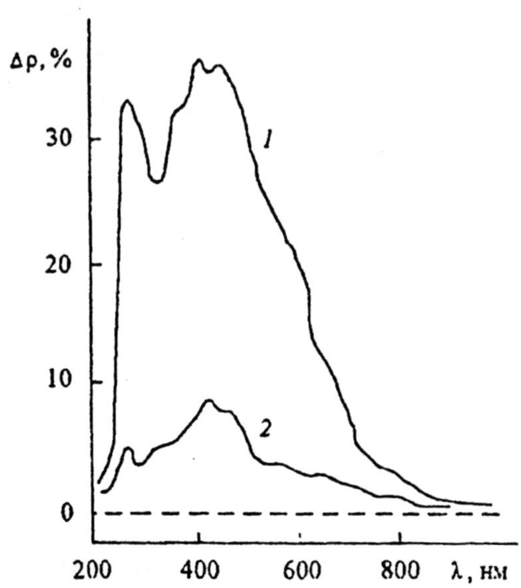

Fig.5 Difference spectra of diffusion reflection of powder ZrO_2 in initial state (1) and with adding 2% of SiO_2 (2) exposed to solar light of dose of 170 e.s.d.

ZrO_2 and SiO_2 powders have great enough specific surface (S_{sp}=4-6 m^2/g for OSC 9-2 ZrO_2 [12]) and they adsorb water well, resulting the producing of OH^- group on the surface. During the exposure, the dehydration occurs, which can be presented as reaction [14]

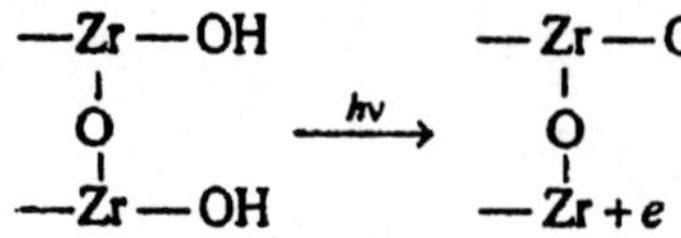

(6)

with creating the oxygen vacancies, the slightly bound oxygen ions and the zirconium ions. The capturing of free electrons by oxygen vacancies, creation of which is possible during ionization of both the matrix and the traps, leads to appearance of F and F'-centers.

Photolysis process occurs with creating the pair: zirconium ions and free electrons

$$Zr \xrightarrow{h\nu} Zr^{2+} + 2e \qquad (7)$$

Mobile electrons are captured by defects with creating unstable radicals

$$[ZrO] + 2e \quad [ZrO^{2-}] \qquad (8)$$

holes are created near valence band

$$[Zr] \quad Zr^{2+} + O^{2-} \qquad (9)$$

they are mobile and can be captured by acceptors, which are the silicon dioxide in a given case

$$SiO2 + O^{2-} \quad SiO_3^{2-} \qquad (10)$$

Annihilation reaction can be written, using equation (7) and (10)

$$Zr^{2+} + SiO_3^{2-} \quad [ZrSiO_3] \qquad (11)$$

Charge-unstable complex is formed, in which the shift of one of oxygen atom towards three-coordinated silicon SiO_3^{2-} occurs.

Under the deeper ionization of zirconium up to Zr^{4+} there is possible the reaction:

$$Zr^{4+} + SiO_3^{2-} \quad ZrSiO_3^{2+}, \qquad (12)$$

the capture of hole leads to creation of stable molecule

$$ZrSiO_3^{2+} + O^{2-} \quad ZrSiO_4. \qquad (13)$$

On the base of curve DTA (Fig.3) one can conclude that interaction between ZrO_2 and SiO_2 is possible only at temperatures higher of 900^0C, the powder exposed to radiation, is heated up to $60\text{-}100^0C$, therefore the reactions (11-12) seem to be hypothetical for photoexcitation. But in spectrum of lamp DKSR-3000, the ultraviolet region is not less 7%. It leads to appearance of activated states: electrons and "hot holes" [14], which are particularly able to rebuild initial crystalline structure of material, that must effect on molecular oscillations and the changes of their symmetry. The comparison of infrared absorption spectra of an exposed and initial samples (Fig.6) shows the disappearance of peak in region of 1400 cm^{-1}, which is typical for ZrO_2, that can be the evidence of such an assumptions.

Thus, experimental results and their analysis show that during preliminary sintering of zirconium dioxide grains, the additional areas of interactions between ZrO_2 and SiO_2 can appear, which assist in increasing a number of optical contacts and decreasing the light scattering near boundaries. Such contacts cause the appearance of intermediate metastable structures during photolysis, which take away the part of incident energy that increases the resistance of coatings to solar light.

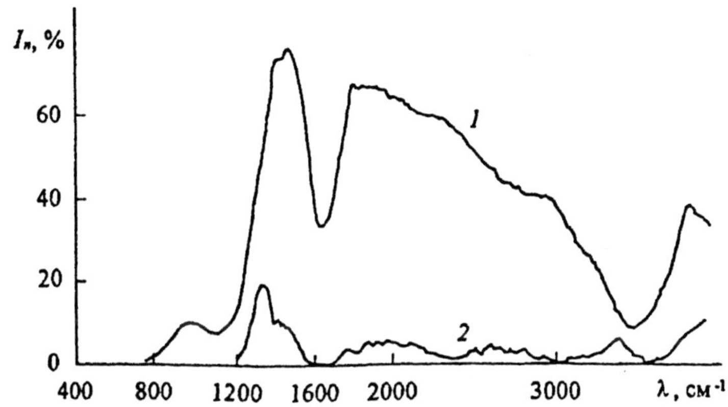

Fig.6 Infrared absorption spectra of ZrO_2+2%SiO_2 prior (1) and after (2) solar light exposure of dose of 170 e.s.d.

CONCLUSION

1. The pigment on the base of ZrO_2 modified by SiO_2 has been developed and the coating has been obtained, which corresponds to reflecting ability of thermoregulating coatings of "solar reflectors" class, which are more resistant to electromagnetic radiation of spectral distribution close to solar one.
2. The analysis of diffusion reflection spectra and infrared absorption spectra has been carried out, the decrease of absorption in wide region of infrared spectra during exposure of coating to solar light has been exhibited.

Authors are grateful to Gorbacheva V.V for preparing the coating samples

REFERENCES

1. Mikhaylov M.M., Dvoretskiy M.I., Kuznetsov N.Ya. Izv.AN SSSR Neorgan, materialy, 1984, t.20, no.3, s.449-453.
2. Emmet J.L., Krupke U.F., Trepholli J.B. Kvantovaya elektronika, 1983, t.10, no.1, s.5-44.
3. Rempel O.I., Driker B.N., Rutman D.S. i dr. Sposob polucheniya stabilizirovannoy dvuokisi tsirkoniya. A.S. no.522138.
4. Savelev GG., Ivanov G.F., Stas N.F. Izv. AN SSSR. Neorgan, materialy, 1988, t.24, no.6, s.960-963.

5. Mikhaylov M.M., Kuznetsov N.Ya., Stas N.F., Dvoretskiy M.I., Aryanov A.P., Gorbacheva V.V., Pogrebenkova, GA. Izv. AN SSSR. Neorgan. Mate-rialy, 1990, t.26, no.9, s.1889-1892.
6. Landsberg GS. Optika. M.: Nauka, 1976, 928 s.
7. Belenkiy E.F., Riskin I.V. Khimiya i tehnologiya pigmentov. L.: Himiya, 1974, 635 s.
8. Milton P.B., Vejnoska L.W., Frederick M.J. Off Digest, 1961,v.33,no.441,p.l264.
9. Mikhaylov M.M., Kuznetsov N.Ya., Ryabchikova L.E. Izv. AN SSSR, Neorgan. Materialy, 1988, t.24, no.7, s. 1136-1140.
10. Kositsyn L.G., Mihaylov M.M., Kuznetsov N.Ya., Dvoretskiy M.I. PTE, 1985, no.4, s.1976-1980.
11. Van den Brom W.E., Volger J. Physica, 1974, v.75, no.2, p.245-246.
12. Mikhaylov M.M., Kuznetsov N.Ya., Ryabchikova L.E. Izv. AN SSSR, Neorgan. Materialy, 1985, t.21, no.2, s.265-268.
13. Polezhaev Yu.N., Kortov B.C., Mishkevich M.V. Izv. AN SSSR, Neorgan. Materialy, 1975, t.11, no.3, s.486-490.
14. Savielev GG., Vladimirov V.M., Mikhaylov M.M., Dvoretskiy M.I. Himiya tverdogo sostoyaniya. Kemerovo, Kem-GU, 1980,s.144-151.

About the Authors

***Vladimirov Valerii Mikhailovich** – Tomsk Polytechnic University, Candidate of Technical science, post doctor. An expert in area of ceramics, chemistry and physics of solid state.*

Mikhailov Mikhail Mikhailovich - *Tomsk Polytechnic University. Professor, Doctor of Technical Science, head of Chair. An expert in the area of radiation physics, radiation materials science and materials for aerospacecrafts.*

OBTAINING MULTILAYER COMPOSITES ON THE BASIS OF AGE-HARDENED AND DISPERSION-HARDENING ALUMINIUM ALLOYS REINFORCED BY "CULON" CARBON TAPES

A.N. Varenkov, V.I.Kostikov, M.G.Krashennikov and N.M.Komarova

The technology for production and methods of evaluation of the quality of carbon-aluminum composite materials (CACM) on the basis of aluminum matrixes reinforced by "Culon" carbon tapes have been developed. Aluminum alloys of specification B-95 or AD33, containing additionally up to 20 vol.% of TiC or SiC, have been used as matric alloys. The effect of heat treatment, shaping and reinforcement of fibers of matric alloys on composite strength has been studied. Use of new matric additions in the composites allowed to obtain new carbon-aluminum material of a high strength and modulus of elasticity. The obtained CACM are recommended to use in productions of tail components parts of aeronautical engineering. It has been shown that use of matric aluminum alloys B-95 containing 2 mass.% of Ti and AD33 containing up to 20 vol.% of Ti or SiC and carbon reinforcing carbon builder on the base of "CULON" is prospective for creating new composite constructional materials for aeronautical engineering.

INTRODUCTION

In fibrous composite materials, the correctly chosen technological scheme and the keeping of regimes of their production causes essential influence on forming the structure and composites properties. Hence, one should to proceed from these conditions:

- technological process must provide uniform distribution of fibers and fibers must be isolated each from other by matrix layer;
- mechanical damage of fibers must be minimal one;
- interaction of fibers with the surrounding and with matrix during their matching and the decrease of fiber strength

must be minimal.

The aim of using the high-strength carbon fibers as builder of composition materials on the base of aluminum, is to obtain high indices of composites qualities, their strength. Combination of low density of armature (up to 1820 kg/m^3) and matrix (up to 2700 kg/m^3) [1]

allows to create the metal constructional composites of high specific strength and modulus of elasticity.

TECHNOLOGICAL EQUIPMENT, PARAMETERS AND REGIMES FOR PRODUCING THE LAYERED CACM ON THE BASE OF AGE-HARDENED AND DISPERSION-HARDENING ALUMINIUM MATRIX ALLOYS

For obtaining the multilayer composite materials, the direct heating plant URK-2 (Moscow State Institute of Steels and Alloys's construction) was employed.

The components of sandwich CACM were obtained in the following way. By means of liquid-phase technology technique the carbon tape "CULON" by was impregnated with melt on the base of aluminium B-95, thus obtaining "prepregs". The pressure of impregnation during the contact interaction between components of "prepreg" was decreased to $20 \cdot 10^5$ Pa. The foil obtained by means of flattening the sheets on the base of age-hardened alloy B-95 (containing 2 mass.% Ti) or on the base of alloy AD33 hardened by dispersion particles of SiC or TiC on rolling mill 250, was used as matrix component. The foil of thickness 0.2 mm and 0.17-0.2 mm, respectively, has been obtained.

The production of multilayer carbon-aluminium composite includes the following technological operations:

1. Assembling the "parcel" consisting of sequentially alternating layers of "prepreg" and matrix aluminium alloy on the base of B-95 or AD33, containing up to 20 vol.% TiC or SiC. The alloy mass is given from experimental data for expansion.
2. The placing of "parcel" between two thin graphite heaters in vacuum furnace, as shown in Fig.1, provided the uniform heating from both sides of "parcel" and excluded the possibility of "swelling" during the heating to the point of melting the matrix alloy foil. The shrinking of "assembly" is carried out by means of rigidity of graphite heaters, the ends of which were fixed in special holders, allowing varying the effort, applied to the heater plates.
3. The heating of "parcel" is carried out to temperatures 800 and 850^{0}C, at which the wetting occurs in system "prepreg"-B-95 and in system "prepreg"-(AD33+TiC), respectively. The aging of composite components under isothermal conditions, with taking into account a whole time of expansion, is not more than 30 min.
4. The cooling of the heating system of furnace and removement of CACM package preform from furnace for the following heat treatment (annealing) is carried out.

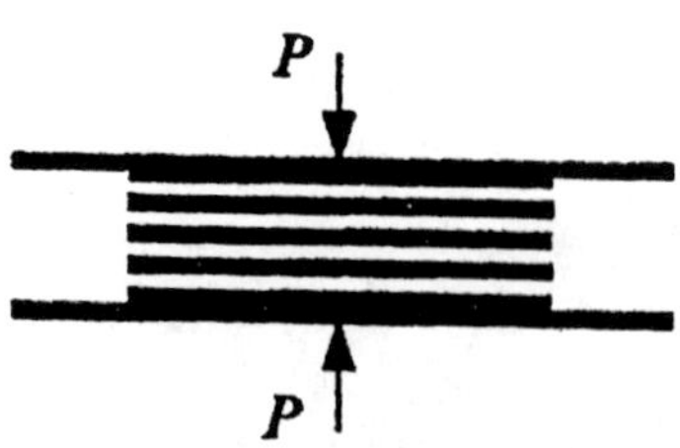

Fig. 1 "Parcel" arrangement (P – applied effort to the sample).

For obtaining the composite materials from "prepreg" with B-95 matrix foil (Fig.2a), or from "prepreg" with AD33+20 vol.% TiC (20 vol.% SiC) foil (Fig.2b), the hot-pressing of assembly was used in vacuum. After that, the straightening of sheet half-finished material CACM by heating the composite for shrinkage on the air was carryied out. The shrinkage was accomplished on the press PR6120 no.3-91 of effort of 10 tons, at ε_{ON}=10%, P_{sp}=200 kg/cm^2, at ε_{ON}=20%, P_{sp}=400 kg/cm^2.

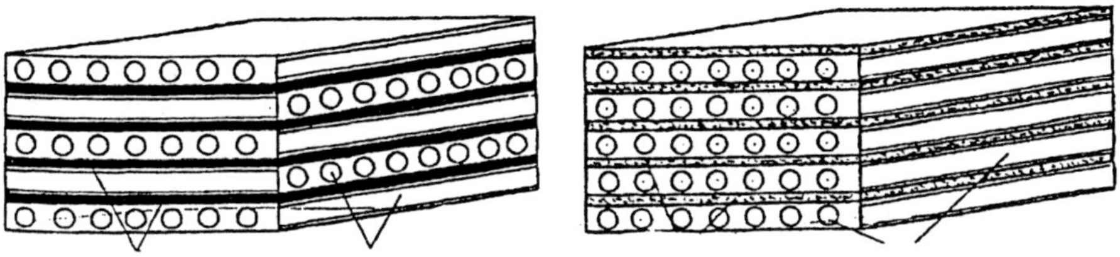

Fig.2 Samples models of CACM: a – two-dimension-reinforced carbonaluminium composition material on the base of dispersion-hardening aluminium alloy V-95+2 of Ti mass.%; b – one-dimension-reinforced carbonaluminium composite on the base of dispersion-hardening matrix alloy AD33, containing 18-20 vol.% TiC or 18-20 vol.% SiC).

PHASE COMPOSITION OF ALUMINUM ALLOYS USED AS MATRIX COMPONENTS IN LAYER CACM

Matrix components on the base of alloys B-95 and AD33, which are used, are of a complex chemical composition and are of four-six-component ones (Table 1).

Multicomponent system causes the appearance of complex hardening phases as the result of heat-treatment used for maximum hardening the materials [2]. The appearance of phases $Al_2Zn_3Mg_3$ and $MgZn_2$ is possible in alloy B-95, being excessive ones. The formation of intermediate phase Al2CuMg follows the doping of alloys B-95 with copper.

When the content of silicon is more than 0.06 mass.% the magnesium silicide $MgSi_2$ is revealed in alloy B-95.

The main hardening phase in AD33 alloys is phase Mg_2Si. Thus the increase of amount of magnesium in alloys up to 1.2-1.4% gives, respectively, the increase of hardening effect and the effect of the aging [3].

The alloys of system Al-Mg-Zn-Cu (B-95) and Al-Mg-Si (AD33) are thermally treated alloys, which are aging-able at both room and higher temperatures. These alloys are characterized by wide range of temperatures enough at which one can carry out the hardening heating. The peak of the strength and yield point is achieved in the case of heat hardening starting from temperatures 470-485^{0}C. The cooling during the hardening is sharp, and the time prior to the dipping into water after withdrawal from furnace is minimal.

The additional doping of B-95 and AD33 aluminium alloys (Table 1) with carbide-forming elements Ti and Zr allows to make these alloys adhesion-active and provides the best conditions of surface wetting of carbon fiber building agent with creating the carbon fibers of carbides of respective materials. However, the introduction of the doping element Ti or Zr into aluminum alloys increases sharply the viscosity of matrix element, therefore its amount must not exceed 2 mass.%. Other doping elements, which are in alloy B-95 and AD33, effect on viscosity slightly [4].

Table 1: Chemical composition of high-strength alloys on the base of aluminium

Broad of the Alloy	Mass Share of Additive					
	Zn	Mg	Cu	Mn	Cr	Ti, Zr, Si
B-95	5.0-7.0	1.8-2.8	1.4-2.0	0.2-0.6	0.1-0.25	1 or 2 Zr
B-96	7.6-8.6	2.5-3.2	2.2-2.8	0.2-0.5	0.1-0.25	1 or 2 Ti
AD33	—	0.8-1.2	0.2-0.4	—	0.15-0.35	1 or 2 Si
AD33	—	0.8-1.2	0.2-0.4	—	0.15-0.35	1 or 2 Ti

The properties of alloys after heat-treatment (hardening at 480^{0}C in water, aging at 180^{0}C during 6-8 hours) are: $\sigma_B^{B\text{-}95}$=550MPa, $\sigma_{0.2}^{B\text{-}95}$=480MPa, σ_B^{AD33}=130MPa, $\sigma_B^{AD33+SiC}$=204MPa, $\sigma_B^{AD33+TiC}$=420Mpa

Structural Fault Probability of CACM After Different Deformation Regimes

Use of dispersion-hardening alloy B-95 of heightened strength, undergone to heat treatment during the obtaining of CACM and reinforced by carbonic fibers of "CULON" tape, allows to obtain ultimate strength up to 920 MPa (Fig.3, curve 1) when introducing 2 mass percentage of titanium into matrix melt B-95. The properties of CACM were obtained by means of the testing to plastometer. In Fig.3, the CACM properties are presented. It has been shown that different kinds of treatment by pressure (rolling, yielding) in the case of using the alloy B-95 as matrix component, without introducing 2 mass.% of Ti into it, does not provide the possibility of obtaining a higher level of properties, hense σ_B reaches only 900 MPa(Fig.3, curve II).

The ultimate strength of CACM corresponds to optimal values of this parameter for different kinds of CACM treatment.

The investigations were carried out on samples CACM, which were deformed at relative height of cogging Δh=0.5 and 1 mm under orthotrophic reinforcement and heating temperatures for rolling of 400-650^{0}C with step of 50^{0}C; the best result responds to point 1 on the curve II, Fig.3.

The structure of samples under orthotropic and two-dimension reinforcement of CACM matrix after having been cogged under the yielding with height deformation value ε_{ON}=10% and 20% within temperature heating interval of 400-650^{0}C with step of 50^{0}C; the optimal value of temporal resistance corresponds to point 2 on the curve II, Fig.3 (for two-dimension reinforcement) and to point 3 on the curve II, Fig.3 (for orthotropic reinforcement).

Studying the structural imperfections of CACM samples during the rolling and yielding, the heating rate for materials treatment by pressure was constant of 26^{0}C/min.

The best result of strength index – temporal resistance corresponds to point 4 on the curve II, Fig.3, resulting from using the various-rate heating of system carbon-aluminium, with minimal heating rate of 8^{0}C/min, and with maximal heating rate of 750^{0}C/min, within heating temperature interval of furnace of 400-1000^{0}C with step of 100^{0}C. The investigations in this case were carried out with CACM at one-direction reinforcement by carbonic fibers.

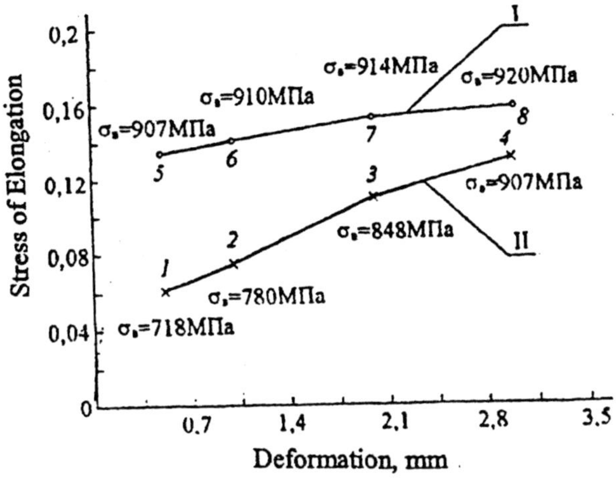

Fig. 3 The parameters of deformation of samples of CACM on the base of dispersion-hardening aluminium matrixes (B-95; B-95+2% Ti), reinforced by "CULON" tapes carbonic fibers (II: 1 – system "CULON"-B-95, the rolling, Dh=0.5 mm, unidirectional reinforcement; 2 – system "CULON"-B-95, the yielding, ε_{ON}=20% two-dimension reinforcement; 3 – system "CULON"-B-95, yielding, ε_{ON}=10%, unidirectional reinforcement; 4 – system "CULON"-B-95, high-rate heating; I: 5 – system "CULON"-B-95, two-dimension reinforcement; 6 – system "CULON"-B-95, unidirectional reinforcement; 7 – system "CULON"-B-95+2% Ti mass.%, two-dimensional reinforcement; 8 – system "CULON"-B-95+2 mass.% Ti, unidirectional reinforcement).

THE EFFECT OF PROCESS CONDITIONS OF CACM ON SPECIFIC CHARACTERISTICS OF COMPOSITES

The numerical values of modulus of elasticity, specific modulus of elasticity and specific ultimate strength of CACM on the base of B-95 alloy, as well as dispersion-hardening matrix alloys on the base of AD33 alloy are presented in Table 2.

According to the data given in table 2, one can conclude that use of discrete particles as hardening phase of matrix metal on the base of AD33 alloy, has provided the possibility of increasing the modulus of elasticity of matrix material up to 128 GPa, whereas for aluminium alloys this index does nor exceed 80 GPa. Specific modulus of elasticity E_{sp} at that is equal to 41830 km, specific ultimate strength of CACM σ_B is equal to 13.7 km. For CACM on the base of dispersion-hardening alloy B-95 the modulus of elasticity increases up to 177 GPa, E_{sp}=80000 km, σ_B=41.2 km. Introducing the adhesion-active Ti element into AD33 alloy, modulus of elasticity reaches 234 GPa, Esp=106000 km, σ_B=42 km.

Table 2: Modulus of elasticity, specific modulus of elasticity and specific ultimate strength of CACM, reinforced by carbonic fibers of tape "CULON" (volume portion v_f=0.53)

System	E, GPa	E_{sp}, km	σ_{sp}, km
"prepreg" ("CULON"+B-95)- B-95+2 mass.% Ti; unidirectional reinforcement	234	106000	42.0
"prepreg" ("CULON"+B-95)- B-95+2 mass.% of Ti, two-dimensional; Reinforcement	218	99090	41.5
"prepreg" ("CULON"+B-95)-B-95 unidirectional reinforcement	209	95000	41.3
"prepreg" ("CULON"+B-95)-B-95 two-dimensional reinforcement	198	90000	41.2
"prepreg" ("CULON"+B-95)-B-95 high-rate heating	177	80000	40.9
"prepreg" ("CULON"+B-95)-B-95 yielding, unidirectional; reinforcement, ε_{ON}=10%	153	69500	38.5
"prepreg" ("CULON"+B-95)-B-95 yielding, two-dimensional reinforcement, ε_{ON}=20%	134	60900	35.5
"prepreg" ("CULON"+B-95)-B-95 rolling, unidirectional; reinforcement, Δh=0.5 mm	119	54090	32.5
AD33+18-20 vol.% of TiC (die component)	128	41830	13.7
AD33+18-20 vol% of SiC (die component)	90.5	3232	7.3

CONCLUSION

1. Technological scheme of obtaining the high-strength compact constructional materials, reinforced by carbonic tape, on the base of composition matrix aluminium alloys has been developed.
2. Technological technique, agreed as the base, which foresees the heating of "parcel" of layer CACM by means of directly passing of current in the process of primary obtaining of the composite at high heating and cooling rate allows to avoid the appearance of aluminuim carbide on phase boundary of CACM components.
3. A carried out evaluation of strength characteristics of CACM at different kinds of composite treatment has provided the optimization of regimes of secondary obtaining of composition materials under conditions of chemical interaction of its components.
4. The regimes of annealing and the regimes (temperature, pressure, deformation) of treatment materials by pressure (rolling, yielding), which excludes the damage of the reinforcing component have been proposed, that has provided the obtaining of the following maximum material characteristics: σ_B=920 MPa, E=234 GPa, σ_{sp}=106000 km, ρ^{CACM}=2200 kg/m^3.
5. An obtained carbon-aluminium constructional composite of high strength under conditions of chemical interaction of materials components are recommended for using in structural units of aeronautical engineering.

REFERENCES

1. 1 Uglerodnye volokna i kompozity. Red. E.Fittser. M.: Mir,1988,345 s.
2. Alyuminievye splavy. Promyshlennye deformiruemye i liteynye splavy. Spravochnoe

rukovodstvo. Red. F.I.Kvasov, I.N.Fridlyander. M.: Metallurgiya, 1984,394 s.
3. Alyuminievye splavy. Vypusk 4. Zharoprochnye i vysokoprochnye splavy. Red. I.N.Fridlyander. M.: Metallurgiya, 197, 396 s.
4. Kompozitsionnye materialy. Red. L.Brautman, R.Krok. t.4. Kompozitsionnye materialy s metallicheskoy matritsey. M.: Mir, 1978, 463 s.

ABOUT THE AUTHORS

Anatolii Nikolaievich Varenkov– *Moscow Institute of Steel and Alloys (Technological University), Doctor of Technical sciences, Professor. An expert in the area of processes and technologies of new multi-functional constructional materials on the basis of graphite, metals and alloys.*

Valerii Ivanovich Kostikov– *Research Institute of Graphite, Doctor of Technical Sciences, Director, Professor. Corresponding Member of RAS. An expert in area of processes and technologies of new multifunctional constructional materials on the basis of graphite, metals, alloys.*

Mikhail Georgievich Krashennikov - *Moscow Institute of Steel and Alloys (Technological University), Candidate of Technical Science, Professor,. An expert in the area of physics-chemistry of metallurgy processes of ferrous metals.*

Natalia Mikhailovna Komarova - *Moscow Institute of Steel and Alloys (Technological University), Candidate of Technical Science,. An expert in the area of structural analysis of multicomponent constructional composites.*

COPPER-ALUMINUM METAL COMPOUND BEHAVIOR UNDER LONG-TERM ION-ELECTRON BOMBARDMENT

G.G.Bondarenko, S.M.Zhdanov, A.P.Korzhavyi and A.N.Tikhonov

The effect of electron and ion bombardment during several thousands hours in vacuum and gas-discharge tubes on the resistance to the sputtering of different emitting compositions on a base of copper and aluminium with oxides has been studied. The dependence of the sputtering resistance vs. the composition, structure and temperature of composition cathodes has been obtained. The presence of lithium oxide in copper and aluminum compositions increases the stability of their properties and provides long-time sputtering resistance.

INTRODUCTION

The processes occurring in bulk and on surface of materials, being the sources of electrons (cathodes) of vacuum and gas discharge tubes, play an important role in providing high reliability and durability of the latter. The emission of electrons from cathodes in tubes of the magnetron type and atomic (helium-neon) lasers is provided, as a rule, at relatively low (300-700 K) temperatures due to electron or ion bombardment. The efficiency of such quasi-cold cathodes, therefore, is considerably determined by values of secondary electron emission coefficients (σ) and ion-electron (γ) emission of the bombarded emitting surface.

In magnetron devices, the vacuum is usually on the level of 10^{-6} mm of mercury (10^{-4} Pa), the energy of primary electrons E_p, bombarding cathode, is of order of 1 keV. In such vacuum the collision of electrons with residual gases in the tubes leads to their ionization by electron impact. Because of that, the cathode is undergone not only to the electron, but also to the ion bombardment. For electron bombardment of metals, the average depth, from which the excited electrons can come into vacuum, is several atomic layers, and σ is not larger 2. Exactly on such depth the defects can be created in metallic cathodes.

Bombardment of the cathodes in vacuum devices, caused by ions produced from ionization of residual gases by electron impact, leads to sputtering their surface. The sputtering coefficient (the number of atoms knocked out from cathode by one ion) depends on impulse energy and ion nature, as well as nature of metallic cathode, its purity and roughness [1,2]. It is quite difficult to pick out the contribution of residual gas ions into electron emission from cathode (ion-electron and ion-ion emission) [3], however it is essential in producing the defects on the bombarded surface [4].

The processes connected with interaction of the ions with metallic cold cathodes in glow discharge tubes in rare gases was studied well enough [5]. The specificity of glow discharge is the inhomogenity of ion flow and complex nature of influence of gas pressure and the parameters of cathode voltage drop on average ions energy. The increase of ions energy higher of some threshold value (usually of 10-30 eV), causes the increase of sputtering coefficient as well, the increase of ion mass effects analogously. The higher evaporation heat of metallic cathode the lesser sputtering coefficient. The average energy of ions hitting the cathode decreases with increasing the gas pressure. The helium-neon tubes of glow discharge, considered in a given work, have a filling pressure of 3 mm of mercury (400 Pa) at partial pressure of $P_{He}:P_{Ne}=5:1-10:1$, the range of discharge current is varied from 2 to 20 mA. The ions energy is defined by cathode voltage drop U_k. The thickness of cathode layer d_k for the considered helium-neon tubes with cold cathode is ~3 cm.

It should be noted that during the exhausting of the gas discharge lasers to $1\cdot10^{-6}$ mm of mercury ($1\cdot10^{-4}$ Pa) and their filling with spectral-pure helium and neon, despite of thorough vacuum-thermal treatment and ion-plasma purifying, the presence of residual atmosphere in them is not excluded. Therefore together with physical sputtering of cold cathodes, the chemical sputtering so far as will appear [5]. Since the copper and aluminium were mostly used as the targets for studying the interaction mechanism of electrons and ions with surface of solid state, the subjects of inquiry were quasi-cold cathodes created on their base.

PROBLEM STATEMENT

The calculations show [6,7] that at the energy of order of 100 eV the runs of ions in solid state do not exceed 2 nm (20 Å), and the atoms sputtering originates basically from first two near-surface atomic layers even at ions energy, exceeding 10 keV [8]. In two-phase alloy, the sputtering coefficient is close to the sputtering coefficient of component difficult to sputter when its concentration is about of 15-20% [9]. It is also known that sputtering coefficient of most the oxides exceeds the sputtering coefficient of metals [10], and that with increasing porosity of substance its sputtering coefficient can decrease [11]. It enables using composite materials of type of metal-dielectric as the electron sources for providing the low sputtering coefficient and, consequently, a high durability for working in both vacuum and gas discharge tubes. Moreover, oxides, possessing a high secondary-emission properties [3], allow sharply increasing efficiency of metallic cathodes with oxide dopant. Although the questions of resistance to electron and ion bombardment of composite quasi-cold cathodes in vacuum and glow discharge, depending on their nature, are studied enough [16,17], the influence of defects, roughness and chemical composition of such a cathodes on emission properties and their stability in time are not unambiguously evaluated yet.

The used in this work devices and methods of scanning electron microscopy allow to study the behavior of cathode surfaces exposed to electrons and ions during long time [12,13], that will allow to determine and recommend optimal regimes of obtaining and exploration of composite quasicold cathodes.

The experimental compositions were obtained by powder metallurgy method when in metallic matrix the oxides are uniformly distributed on surface and bulk , as well as by epitaxial methods when the composition is made of two-layer system – oxide(coating)-metallic base [14]. And, finally, the composition cathodes were obtained as films (coatings) by their deposition on the substrates of necessary configuration and nature [15].

THE FEATURES OF OBTAINING THE COMPOSITE QUASI-COLD CATHODES

The composition cathodes on the base of copper and aluminium have the complicated oxide dopands [16-20]. The distinctive feature of investigated samples is that they all had lithium oxide in themselves composition, providing the high resistance of compositions to bombardment in both vacuum [21] and glow discharge [22].

Bulk Compositions on the Base of Copper

The copper powder of specification PMA by GOST 4960-75 was employed as metallic base, and the mixture $BaO+L_2O$ as oxide component. As far the barium and lithium oxides in pure form are not produced, carbonic barium ($BaCO_3$) of specification HC by GOST 4158-72 and carbonic lithium Li_2CO_3 of specification OSC 19-2 by TU 6-06-282-72 were used. Barium and lithium carbonates, mixed in necessary compositions and pressed in cylindrical billet, were heated in vacuum for choosing optimal regimes of obtaining the oxides. The data of X-ray analysis of mixtures $BaCO_3+Li_2CO_3$ are given in Table 1.

Table 1 The dependence of phase composition of mixtures $BaCO_3+Li_2CO_3$ vs. annealing regimes.

Annealing regime at $P=10^{-3}$-10^{-4} Pa		Phase contents
Temperature T, K	Time t, hours	
950	2	$BaCo_3$, Li_2Co_3, traces $Ba(OH)_2$, Li_2O, BaO
	3.5	$BaCo_3$, Li_2Co_3, BaO, Li_2O
	4.5	—"—
	7	—"—
970	4	BaO, Li_2O, $BaCo_3$, Li_2Co_3
1070	2	BaO, Li_2O, $BaCo_3$, Li_2Co_3
	3	BaO, Li_2O traces $BaCo_3$ traces Li_2Co_3
1120	1	BaO, Li_2O

For obtaining the mixtures of barium and lithium oxides, the latter regime was chosen. After heat treatment, the billets were milled in vibrating mill and were sifted. The dimension of obtained particles was less of 15 μm. The initial particles are co-mixed with copper powder and additionally milled by tungsten balls during 10 hours.

The button weight of mixture prior to the pressing was calculated due to equation:

$$P = dV\left(1 - \frac{\Pi}{100}\right)k$$

where P is mixture mass, d is material density, V is volume of the preform, П is the expected porosity, k is coefficient taking into account the mixture loss during the pressing.

Fitted experimental pressure of the pressing was $9.8\text{-}9.9 \cdot 10^{10}$ N/m^2, relative preform density was 75-80%.

For composition the copper with additions of BaO:L_2O at molar ratio 1:1, at total amount of oxides 2-10% in mass, the optimal temperature of sintering in vacuum is within 1150-1170 K. After sintering, the cathode density is 90-95% of theoretical, i.e. their porosity is 5-10%.

Copper-Oxides Film Compositions

The film samples Cu+10-20% BaO-Li_2O on copper and molybdenum substrates were obtained by method of ion-plasmous sputtering of targets manufactured according to scheme mentioned above for bulk compositions. The film systems were obtained in argon medium at pressure of $5 \cdot 10^{-1}\text{-}5 \cdot 10^{-2}$ Pa and voltage of 1.5-2.5 kV, the coating deposition rate is 6-10 nm/min and substrate temperature is 420-500 K. All the samples, independent of their type, prior to experiments were exposed to the activating annealing in vacuum at different temperatures.

Compositions on the Base of Aluminum

Substrates of such cathodes were obtained by mechanical treatment methods from aluminium rods or pipes. After techno-chemical clearing on substrate surface, the oxide films of thickness 1080 nm at oxygen pressure 0.5-5 mm of mercury (67-670 Pa) in temperature range 573-700 K during 0.5-1.5 hours have been grown. The special devices which provide induction heating of the samples with the external electromagnetic field frequency of 0.5 kHz [23] have been used. Oxide film on aluminium substrate consists of two layers. Internal layer is compact and its thickness depends on temperature of the obtaining. The external layer is porous and consists of hydrated Al_2O_3. The thickness of such layer under usual conditions is 10-20 A [6,20].

Experimental Investigation of Resistance of Quasi-Cold Cathodes to Bombardment

Bombardment Resistance in Vacuum. Technique and Apparatus

Construction of experimental tube for determining σ of studied cathodes – targets was typical one [19]. The main modules of tube are: electron gun and "collector-target" are placed in glass balloon. Electron cathode gun is the tungsten thread of diameter of 0.2 mm. Two diaphragms placed before collector are serving for collimating the electron beam and for the work of gun. First diaphragm is under cathode potential, second is under target potential. The metallic screen is also applied, which serve for avoiding the electrons to be hit on collector. The shield is placed under target potential.

The tube module "collector-target" consists of collector of secondary electrons, the target of investigated samples with thermocouple serving for changing the target temperature. To avoid the material sputtering from target and onto target during high-temperature treatment of tube and target modules, the movable damper plate is mounted between target and collector. The target is mounted on tantalum core, inside which the special heater is mounted. The tube contains modulator and electrostatic lens destined for forming the electrons beam and changing their energy. For keeping high vacuum in tube the titanium pump is placed inside it.

Prior to measurements the tube was degassed by heating. The gun cathode, target heater and titanium pump were degassed by passing current. The modules of gun and measuring block were degassed by high frequency currents. The degassing regime of investigated gun is: heating up to 970-1070 K, the pressure in tube, measured by manometric tube PMI, did not exceed $6.6 \cdot 10^{-5}$ Pa. After welding, the pressure in tube was not more than $11.3 \cdot 10^{-5}$ Pa.

The regime of heating treatment of cathode-target and change of pressure of residual gases in experimental device are shown in Fig.1. After complete degassing of the modules, the cathode was activated by the heating in optimal regime (920 K, 1 hour).

Such a construction allows to obtain the dependence of σ vs. electron energy within range from 200 eV to 1.5 keV at primary current of 10^{-2} A.

The electron beam diameter was able changing from $1 \cdot 10^3$ to $5 \cdot 10^3$ μm.

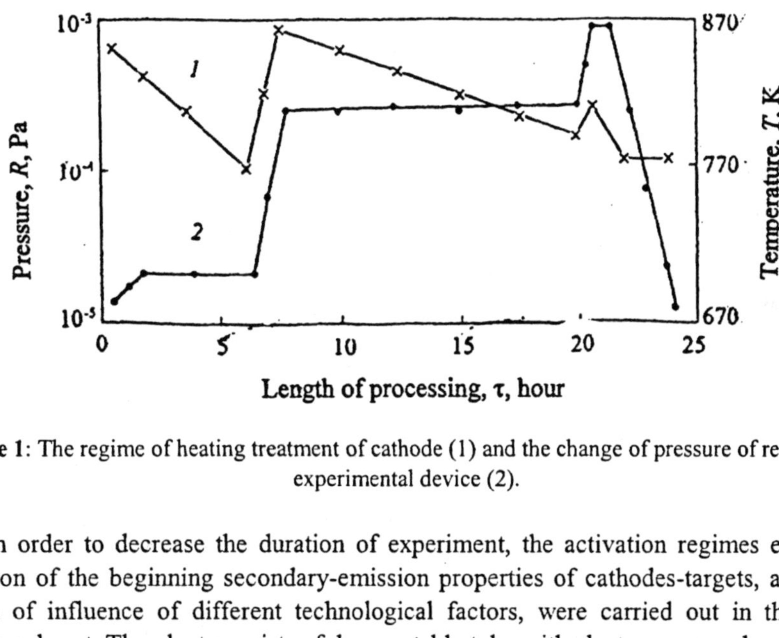

Figure 1: The regime of heating treatment of cathode (1) and the change of pressure of residual gases in experimental device (2).

In order to decrease the duration of experiment, the activation regimes execution and investigation of the beginning secondary-emission properties of cathodes-targets, as well as the evaluation of influence of different technological factors, were carried out in the plant with continuous exhaust. The plant consists of demountable tube with electron gun and system exhaust. The construction of demountable device does not principally differ of construction of experimental tubes. A small changes were introduced into it, allowing to determine the secondary electron emission coefficient (SEEC) and first critical potential E_p^1 beginning from value of 10 eV with accuracy of ±2 eV. While measuring σ, the exhausting system was provided by means of aggregate ERA-100, the pressure of residual gases in device was $1.3 \cdot 10^{-5}$ Pa. The measurement circuit and the feed circuit of device modules are analogous to the circuits of the tube device. Measurement of σ on a given plant was carried out in static regime.

The investigation of dissociation products extracted during electron beam exposure of composition cathodes-targets was carried out on the plant described in [24].

The investigated composition samples of system Cu-BaO-Li_2O had the microstructure with properties analogous to a shown in Fig.2. For providing the necessary emission activity, the cathode must be exposed to activation process at temperatures exceeding the working temperatures [21].

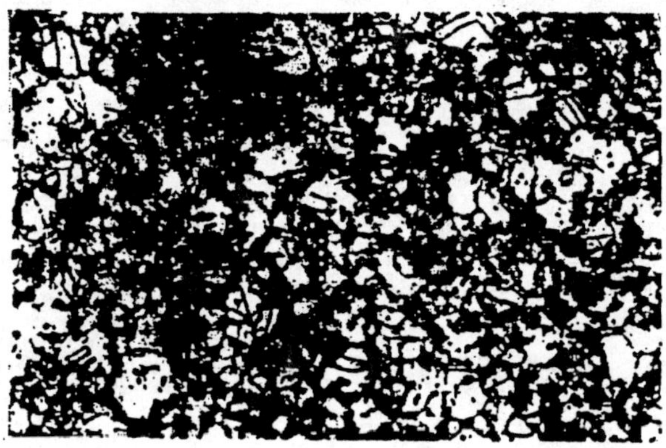

Figure 2 Microstructure of material Cu-BaO-Li_2O after sintering. Magnification, ×340.

The choice of activation regime of experimental samples was carried out in dependence of σ vs. temperature and expose duration. The change of temperature was carried out in discrete steps – trough every 100^0, the exposure during 0.5 hours was made. When plotting the dependence of vs. bombardment electron energy, the target temperature was decreased to 670 K for decreasing the influence of current, and then was increased to necessary one. During heating of the samples, σ increases firstly and then begins decreasing. It has been noticed that activation regime is defined by active addition. Simultaneously with changing σ, value E_{pI}^{I} changes as well (Fig.3). With increasing the temperature, the value E_{pI}^{I} decreases in the beginning, and then the increase is observed, i.e. for every materials there is the activation temperature, when heating to which the first critical potential achieves minimal value.

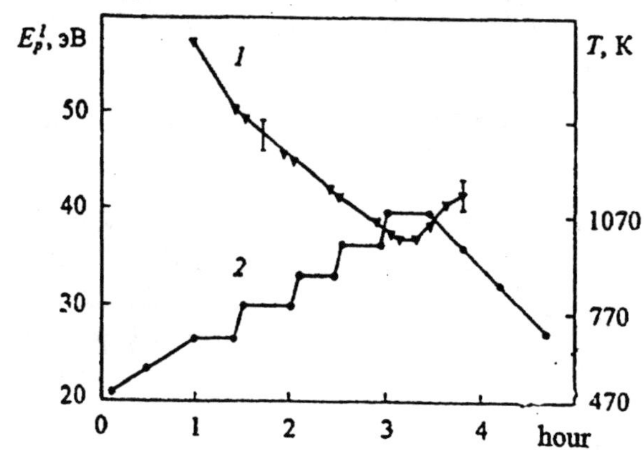

Figure 3 The change of first critical potential (1) and temperature (2) during activation of samples Cu-6% (BaO: Li_2O=1:1).

The dependence of σ vs. activation duration at difference temperatures of samples of composition Cu-6% (BaO:Li_2O=1:1) is presented in Fig.4. The peak value σ, being equal to 3.5, was obtained after having heated the samples at 970 K during 1 hour. The decrease of activation temperature leads to the increase of heating duration necessary for obtaining σ close to 3.5. The increase of activation temperature up to 1070 K leads to quicker increase of σ, but this value appears to be smaller (3-3,1). Analogous behaviors and values of σ and E_p^I during activation are observed for all the investigated cathode materials.

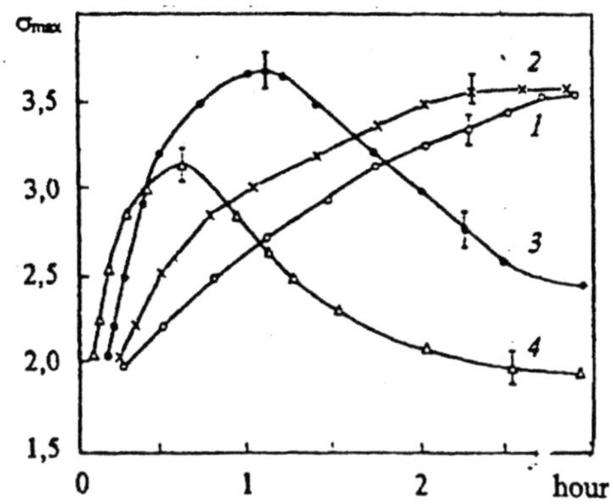

Figure 4 The dependence of secondary electron emission coefficient of samples of contents Cu–6%(BaO:Li_2O=1:1) vs. the duration of activation at temperatures: 1 – 670K, 2 – 770K, 3 – 970K, 4 – 1070K.

The cathode can be exposed to air during different time in the process of the manufacturing and mounting that lead to the change of emission properties. Analysis of the dependence of σ vs. activation temperature of samples, having been exposed to air, shows that peak value of SEEC after activation is close to the values of initial samples, but the activation process is accomplished for shorter time.

One of the reasons of more intensive process of activation after exposure of cathode-target to the air is the forming of the hydrates and carbonates from oxides under influence of CO_2 and H_2O, having the melting temperature lower of activation temperature. For example, for $Ba(OH)_2$ the melting temperature is close to 970 K, for LiOH the temperature is close 715 K. During the heating process the hydrates and carbonates are melted, which are further decomposed, contributing into cathodes gas disengagement. The peak disengagement rate of H_2O is observed at temperature 700-720 K. Within range of temperatures 500-750 K, CO_2 intensively disengages. For CO+N_2, two peaks in region of temperatures of 620K and 870K were observed. On further temperature increasing (up to 1200K) CO+N_2 is appreciably segregating. After exposing at temperature of 920 K during one hour the gas segregation from cathode does not completely stop, but is on the level close to the sensitivity of plant for gas segregation determination [24].

The other processes, occurring during activation, can be evaluated by character of the emitting surface of cathode-target and its change by means of scanning electron microscope MPEM-200. The influence of the heating temperate on the character of emitting surface was investigated at temperatures of 670 and 770 K (Fig.5). It should be noted that during activation, the emission picture change was as the following. At the first, the single emission centers were observed and their activity have grown. Then the glow intensity of these centers relaxed.

The bright spots, observed in emission microscope on the preliminary stage of activation process, connected with places of oxide grains yield, increase gradually themselves area. This fact shows that during activation, the oxides migration on the cathode surface occurs, leading to the increase of degree of surfacing it with oxide films.

The presence of oxides migration over cathode surface is confirmed by data of micro-X-ray analyzer MS-46 by firm "CAMECA". The surface prior to activation is the mosaic of oxides inclusions in metallic matrix. The inclusions have dimension from 1 to 7 μm and are uniformly distributed on surface.

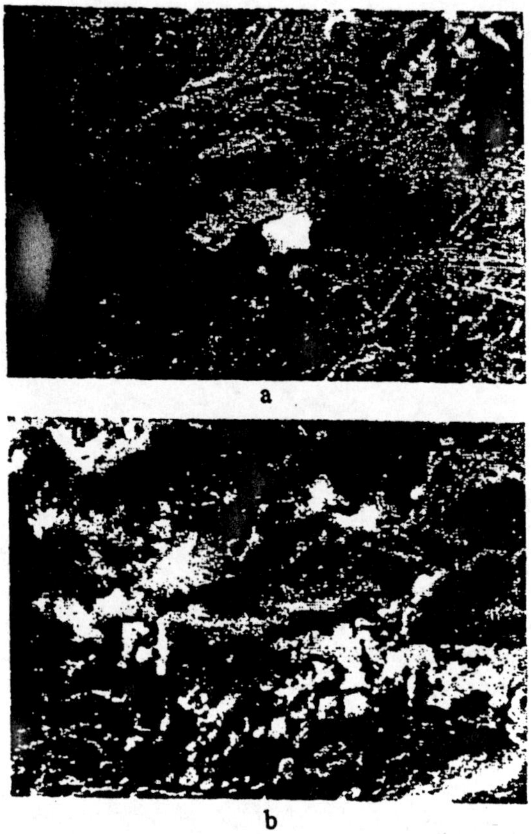

Figure 5 The view of cathode emitting surface after activation at temperatures: a – 670K, b – 70K. Magnification, ×2000.

The warming-up of sample causes essential changes of surface. The increase of the surfacing of metallic matrix by oxide inclusions by means of migration and diffusion of oxides from cathode-target bulk is observed. The dimensions of the oxide buildings on the surface increase to a hundred microns, the boundaries are becoming fuzzy, the barium amount decreases to the edges. The degree of the surface covering by active oxide inclusions depends on the duration of the heating of sample. The migration rate of barium oxides over surface (the rate of film area increasing), obtained by calculation from investigation results at 970 K is approximately equal to $1 \cdot 10^{-2}$ μm/s. The increase of degree of the covering by oxide films of cathode-target working surface, as well as the increase of oxide outlet centers number on the surface, leads to the increasing of its uniformity on thermoemission characteristics.

Some behaviors of the chane of σ of samples during activation of cathode-target can be explained by outlet and migration of oxides, being in its composition, over the working surface. There is optimal ratio and amount of oxides in material at which the value σ is maximal, and the value E_p^I is minimal. For the investigated compositions the amount of oxides of 6-8% in mass at molar ratio 1:1 provides the best emission properties.

The dependence of σ vs. energy of preliminary electrons E_p for materials on the base of copper with additions 6 mass.% of $BaO:Li_2O=1:1$ is given in Fig.6. The peak value of σ is in range of 3.7-3.9 at value of E_p^{max} being equal 400-800 eV. When the concentration of oxides is less than 3 mass.%, a low reproducibility of emission properties is observed. When the concentration is more than 7%, the decrease of σ during investigation in bombardment static regime is observed because of the recharge effect of material surface. The density distribution for samples of one group does not exceed ±3%; σ is differed for 0.1-0.15% towards the decrease when density decreases. The change of σ vs. samples density is defined by scattering the part of secondary electrons in micropores, leading to impossibility their getting off on the surface and increase of σ [12].

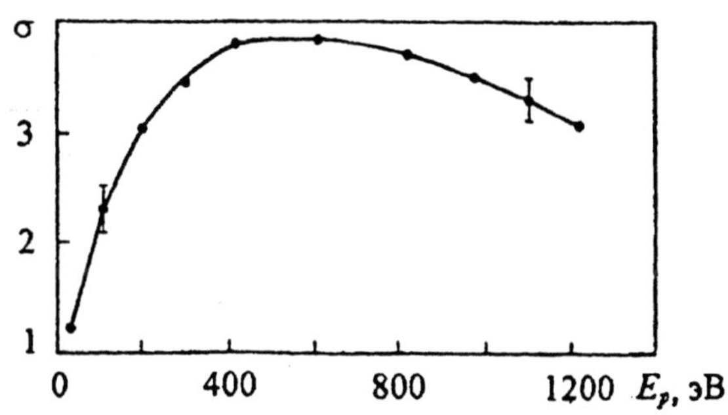

Figure 6 The dependence of $\sigma(E_p)$ for samples Cu-6%($Ba:Li_2O=1:1$).

The distribution of amount of oxides in the samples of one group can lead to changing the value σ within ±0.1%.

The change of σ of investigated samples is determined also by oxides ratio at near-surface layer. The analysis of results obtained for materials with equal total oxides amount, but at 4 different molar ratios shows that difference in σ can reach 1.35, and 15 eV for E_p^I.

Thus, the presented results shows that composition cathodes on the base of metals with additions of oxides mixture have a small distribution of emission properties due to the change of porosity and composition, as well as the molar ratio of oxides in near-surface area and surface films.

Stability of secondary-emission characteristics of cathode materials under effect of electron bombardment, reaching hundreds A/m^2, is one of the basic factors, which define the efficiency and durability of device.

Long-time tests of investigated cathode materials under conditions of exposure of electron flows of different density were carried out on specially manufactured testing benches, where 10 experimental devices were simultaneously set. The handling and the controlling of every device is autonomous.

The dependence of σ vs. bombardment electron energy at different testing regimes was determined periodically, the primary electron beam current short-timely decreased to $5 \cdot 10^{-6}$ A for the period of plotting the dependence, then the testing continued to a given regime.

The change of σ of emitter of composition Cu-6% ($BaO:Li_2O=1:1$) during the testing for 2 thousands hours at temperature 770 K is presented in Fig.7. The increase of primary beam current density leads to decreasing the value σ on stable region of cathode working. The change of σ after long-term testing during different terms and under different regimes is presented in Table 2.

Table 2 The change of secondary-emission characteristics of material after testing in different regimes

Testing regime			Secondary-emission characteristics				Testing duration i, hours
			Initial		Final		
j 10^2, A/m^2	T, K	Ep, keV	σ_{max}	E^I_p, eV	σ_{max}	E^I_p, keV	μ
1	370	1	3.6	50	2.5	60	26500
3	370	1	3.4	50	2.0	65	9800
3	470	1	3.6	35	2.0	65	3500
3	670	1	3.5	40	2.0	70	5200
2	670	1	3.4	40	2.8	55	7700
2	770	1	3.3	30	2.0	65	8200
6	770	1	3.5	30	2.3	60	5600
10	770	1	3.6	35	2.0	70	2500
20	770	1	3.5	35	2.0	75	1000

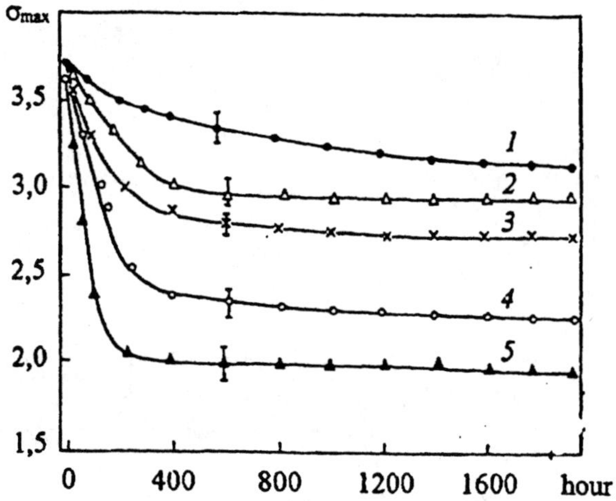

Figure 7 The dependencies of peak value of secondary electron emission coefficient of emitters Cu-Ba-Li_2O vs. bombardment duration by electrons of energy of 1.0 keV and beam density: 1–j=10^2 A/m^2, 2-j=$2\cdot10^2$ A/m^2, 3-j=$3\cdot10^2$ A/m^2, 4-j=$6\cdot10^2$ A/m^2, 5-j=10^3 A/m^2. The target temperature is 770K.

The analysis of obtained results on investigation of influence of long-term electron bombardment under different regimes of the tests has allowed to define the following behaviors distinctive for emitters on the base of copper with additions of oxides mixture:

a initial values of σ for samples under different testing regimes is 3.5-3.8;

b during long-term testing the value σ decreases. The stage of cathode working during which σ decreases approximately for 25% of initial one is observed. The stage duration is from 50 to 500 hours. The decreasing rate of σ depends on the density of bombardment beam current density. With increasing the current density, a more sharp dropping of σ on the beginning stages of the working of cathode is observed. The time to stable part of the working decreases with increasing the current density. At the density of primary current of $1\cdot10^3$ A/m^2, σ is stabilized for 500 hours, at $1\cdot10^3$ A/m^2 – for 400 hours.

The main reason determining the dependence of σ vs. testing time at different density of electron beam is the damage of oxide films produced on the cathode surface during activation, and then the oxide damage in grains and pores.

The Bombardment Resistance in Glow Discharge. Technique and Apparatus

The resistance of film targets to the ion bombardment in gas-impurity glow discharge was carried out with taking into account that heavy component particles are mainly contributing into the sputtering. The sputtering rate was evaluated due to the change of quartz resonator frequency (Δ_f~Δ_f). Oxide films of first samples series of system Al_2O_3 was aluminum A999 of thickness of 0-50 nm and did not contain a noticeable quantity of impurities, the lithium was noticeable impurity in others ones.

The roughness of surface of all the investigated targets was not worth of R_a=0.16-0.08 μm. The highest resistance are provided by cathodes-targets with surface relief, on which the relief with indentations of order of 20 μm, was formed on the stage of technical treatment of substrates prior to oxidation.

The testing of aluminium cathodes on the servicing term was carried out in diodes (model devices) with working volume of helium-neon mixture of 350 cm^3. The detail of construction of diodes was that the dielectric (glass) cone widening from capillary to cathode was placed in the discharge space between cathode and capillary. The pilot experiments have shown that such a scheme of inserting the discharge into cathode-target provides the most uniform distribution of current over its surface. The decrease of pressure of helium-neon mixture for 20% was the criterion of technical age. The pressure was measured by Pirani manometer. Aluminium cathodes-targets were manufactured of diameter of 40 mm and of length of 200 mm. Up to 100 devices were simultaneously tested. For discharging current of 40 mA, the current density near cathode-target did not exceed 0.2 mA/cm^2. The obtained experimental results on cathode durability are presented in Fig.8.

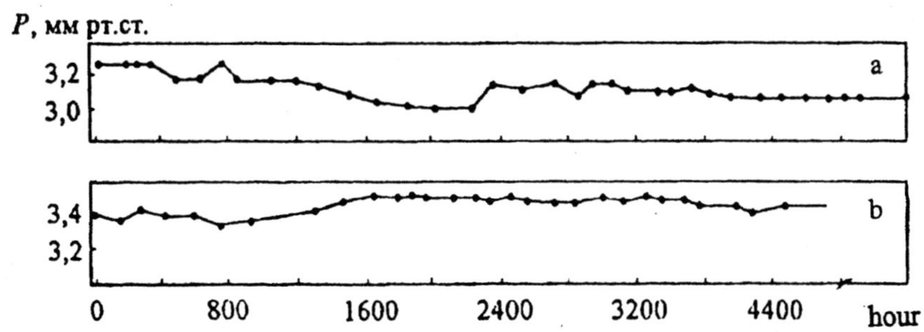

Figure 8 The dependence of the change of total pressere of helium-neon mixture vs. testing time: a – device 38-1,5; b – device 16-1,5.

The first experimental tube device (38-1.5) contained the cathode-target of samples series, manufactured from pure aluminium. The second (16-1.5) had the cathode-target of samples series manufactured from aluminium, containing lithium impurities. The parameters behavior of these devices in the exploitation differ (Fig.8). In the first 500-600 hours, on the stage of stabilization of devices parameters, they were practically equal. After 800 hours, the second device, in which the target (cathode) of oxidized aluminium containing lithium was tested, was stably working, the pressure of gas helium-neon mixture did not decrease in it in comparison with initial (Fig.8b). In testing the first device, the instability depending on pressure of helium-neon mixture vs. bombardment time in analogous regimes of glow discharge was observed. Most likely, the gas absorption mechanism during bombardment of oxidized aluminium target-cathode, contained lithium, is somewhat difference, than for lithium free oxidized aluminium [25]. Probably, during bombardment of quasi-cold target (cathode) preliminary undergone with oxidation, L_2O more easily dissociates, and free lithium participating in gas absorption from tube, because of permanent migration over bombardment surface, promotes in the desorption of absorbed gases. Such a adsorption-desorption mechanism of quasi-cold cathodes, contained lithium, positively influences on devices parameters of glow discharge and promotes in reaching the high durability at relatively high densities of discharge currents of 0.2 mA/sm^2 (in a known foreign gas-discharge helium-neon devices with aluminium cathode the current density is equal to 0.05 mA/sm^2 [26]). It should be noted that cathode-target temperature essentially influences on its durability in the glow discharge.

The investigation of processes occurring in discharge with cold cathode, allowed to clarify the correlation of main physical phenomena and create mathematical model with help of which the analytical expression has been defined, which bind the device durability with cathode parameters in glow discharge. They consider that quantity of gas atoms, leaving the device volume is approximately proportional to quantity of sputtered atoms of cathode and some of them return into cathode, and the cathode durability is inversely proportional to intensity of its sputtering. It has been included also that when passing the current through atomic gas, the heat energy which is proportional to the density of discharge current and electrostatic intensity, is rejected, the cathode voltage drop and the cathode incidence region width change in the near cathode region, because of the largest heating and the decrease of gas concentration. The model is based on measuring the current density J, pressure of gas medium P, gas temperature T. Taking into account that t is cold cathode durability, P_0 and T_0 is pressure and temperature of gas during fulfilling the device, T_k and V_k are temperatures and gas volume in cathode plain, T_A is gas temperature near anode, V is total volume of gas fulfillment, D is constant value for a given sort of gas and cathode material, indices 1 and 2 are forced and working testing regime, respectively, then the forecasting formula is:

$$\frac{t_2}{t_1} = \left(\frac{j_1}{j_2}\right)^2 \left(\frac{P_{0_2} T_{0_1} \delta_1}{P_{0_1} T_{0_2} \delta_2}\right)^3 x$$

$$\exp\left[D\left(\frac{P_{0_2} T_{K_2}}{\sqrt{j_2} T_{0_2} \delta_2} - \frac{P_{0_2} T_{K_1}}{\sqrt{j_1} T_{0_1} \delta_1}\right)\right]$$

where

$$\delta_{1,2} = \frac{2T_{K_{1,2}}}{T_{K_{1,2}} + T_{A_{1,2}}}\left(1 - \frac{V_K}{V}\right) + \frac{V_K}{V}$$

The results of approbation of a proposed forecasting formula of durability of glow discharge with quasicold cathodes are given in Table 3.

Table 3: Results of Testing Aluminium Oxidized Cathodes with Forces and Working Regimes

No/p/p	Test Regime	Density of Current, F_1, mA/cm^2	Pressure, P_o, Pa	Time of Test, L, Ch	Prognosed Durability L, Ch	Deviation from data on direct operating time, %
1	Forced	1.0	400	700	22600	4.5
2	Forced	2.0	400	140	21800	7.2
3	Forced	4.5	400	36	21530	9.1
4	Forced	2.0	266	57	22100	5.9
5	Working	j_2=0.2	P_{o2} = 400	t_2= 235000	—	—

From the theoretical researches and analytical calculations it follows that for all the considered kinds of bombardment of composition aluminium cathodes the general tendency is observed: dependence of resistance to irradiation is connected with the flow density of bombarding particles and temperature.

The physical model of the emission surface structure model has been developed for aluminium oxidized surface of cold cathodes (targets). During the model testing, it has been revealed that when such a surface is the funnel-shaped, recesses of micron depth, the significant part of the sputtered particles does not leave the bulk of these recesses, but precipitate on the walls. Consequently the process of "immuring" the gas into laser details decreases, that increases its durability.

According to data of investigations, the optimal thickness of oxide film on the aluminium is 20 nm (200 Å), since the peak of secondary ion-electron emission coefficient γ is in this region [6].

The typical structure of surface with minimal sputtering degree in the glow discharge (having the assigned relief) is shown in Fig.9 a,b.

In conclusion it should be noted that behavior of quasi-cold cathodes (targets) under bombardment, independent of their performance (either bulk or film composition) is analogous. These experimental results are the principal, since the data of influence of long-term electron and ion bombardment on the emitting composition behavior depending on their structure, composition, temperature and the production techniques have been obtained. The y are important not only from position of radiation materials science but also are of great interest. Non-ferrous metals consumption decreases for 1-2 degree when realizing the film emitting compositions as the quasi-cold cathodes in repetition work in comparison to bulk ones.

a

b

Figure 9 The surface morphology of aluminium cold cathode with the formed recesses on it of micron dimensions for providing the minimal sputtering in glow discharge. The magnification: a - ×300, b -×1000.

CONCLUSION

1. The behaviors of influence of long term electron and ion bombardment on the serviceability of quasi-cold cathodes on the base of aluminium and copper for vacuum and gas-discharge devices have been determined.
2. The optimal conditions of cathode activation have been defined; it has been shown that these conditions depend on not only activation temperature and duration, but also on oxides amount in cathode material.
3. The dependencies of cathode material resistance to bombardment in glow discharge vs. their composition, structure and temperature have been determined. It has been demonstrated that copper and aluminum alloys, containing lithium, have a higher stability of properties.

The given work was carried out with financial supporting of the Russian Foundation of Physical Investigations (RFPI) and the Ministry of Education.

REFERENCES

1. Abroyan I.A., Andronov A.N., Titov A.I. Fizicheskie osnovy elektronnoy i ionnoy tehnologii, 1984, Moscow, Vysshaya shkola, 320 s.
2. M.Kaminskiy. Atomnye i ionnye stolkonoveniya na poverhnosti metalla. Perevod s angl., pod red. akademika L.A. Artsimovicha, 1967, Moscow, Mir, 506 s.
3. Dobretsov L.N., Gomoyunova M.V. Emissionnaya elektronika, 1966, Moscow, Nauka, 564 s.
4. Granovskiy V.L. Elektricheskiy tok v gaze. Ustano-vivshiysya tok, 1971, Moscow,Nauka, s.292-332.
5. Svettsov V.I. Osobennosti raspyleniya razlichnyh materialov pri razryade v himicheski aktivnyh sredah. Obzory po elektronnoy tehnike, ser.4, Elektro-vakuumnye i gazorazryadnye pribory, 1983, Moscow,TsNII "Elektronika", vyp. 5 (979), 40 s.
6. Korzhavyy A.P., Kristya V.I. Fizicheskie protsessy v prikatodnoy oblasti tleyuschego razryada i prognozirovanie dolgovechnosti katodnyh materialov dlya otpayannyh priborov. Obzory po elektronnoy tehnike, ser.6. Materialy, Moscow, TsNII "Elektronika", 1988, vyp.7(1403),ch.1,40s.
7. Biersack J.P, Eckstein W. J.Appl. Phys. A., 1984, v. 34, no.2, p.73-94.
8. Shulga V.I. Poverhnost, 1982, no.3, s. 38-41.
9. Dahlgren S.D,, McClanahen E.D. J.Appl.Phys., 1972, v. 43,no.4,p.l514-1517.
10. Raspylenie tverdyh tel ionnoy bombardirovkoy. 1986, vyp.2, pod red. R.Berisha, Moscow, Mir, s.488.
11. Martynenko T.P. ZhTF, 1968, t.38, no.44, s.759-760.
12. Tihonov A.N., Pankratov V.V. Primenenie avtomatizirovannogo kompleksa REM-EVM dlya tehnologicheskogo kontrolya izdeliy elektronnoy tehniki. Materialy vsesoyuznogo soveschaniya-seminara «Problemy teorii chuvstvitelnosti elektronnyh i elektromehanicheskih ustroystv i sistem», Moscow, Radio i svyaz, 1983,s.51-60.
13. Pankratov V.V., Rybalko V.V., Tihonov A.N. Pribory i tehnika eksperimenta, 1983, no.5, s.200-201.
14. RedegaK.P., Zhdanov S.M., Savranskaya E.S. Elektron-naya tehnika, ser.6. Materialy, 1976, vyp.4, s.24-29.
15. Bondarenko GG., Korzhavyy A.P. Izvestiya RAN, ser. Metally, 1995, no. 4, s.167-171.
16. Zhdanov S.M., FayferS.I.,TarasyukA.I. Elektrorn-naya tehnika, ser.6. Materialy, 1974, no.2, s.20-23.
17. Korzhavy A.R, Journal of Advanced Materials, 1994, no.l,p.46-53.
18. Bryuining G. Fizika i primenenie vtorichnoy elek-tronnoy emissii. Moscow, Sovetskoe radio, 1976, s. 125.
19. Bronshteyn I.M., Frayman B.S. Vtorichnaya elek-tronnaya emissiya. Moscow,Nauka, 1969, 408 s.
20. Korzhavyy A.P., Kristya V.I. Fizicheskie protsessy v prikatodnoy oblasti tleyuschego razryada i progno-zirovanie dolgovechnosti katodnyh materialov dlya otpayannyh priborov. Obzory po elektronnoy tehnike, ser.6. Materialy, Moscow, TsNII "Elektronika", 1989, vyp.4 (1457), ch.P, 40 s.
21. Korzhavyy A.P., Redega K.P, Materialy dlya katodov s nizkimi znacheniyami pervogo kriticheskogo poten-tsiala. Obzory po elektronnoy tehnike, ser.6, Materialy. Moscow, TsNII "Elektronika", 1987, vyp.2, (1269), 39 s.
22. Bondarenko G.G., Korzhavyy A.P., Kristya V.I., Sigov D.N. Izvestiya RAN, ser. Metally, 1993, no.3, s.97-100.

23. A.S. SSSR no. 375709. Sposob izgotovleniya holodnyh metallicheskih katodov. Fayfer S.I, Zhdanov S.M., Korzhavyy A.P., Artsyhovich V.F., Trofimov E.A., Kunets N.N.-B.N, 1973, no.16, s.130.
24. Tarasyuk A.I., Korol L.N., Zhdanov S.M. Elek-tronnaya tehnika, ser. Materialy, 1971, vyp.7, s.23.
25. Ivlev A.M., Korzhavyy A.P., Moskvina A.I. Elek-tronnaya tehnika, ser. Elektrovakuumnye i gazo-razryadnye pribory, 1979, vyp.8, s.67-72.
26. Patent 4101846 SShA. Gas laser / Vichael E. Fein, Charles W. Salisbury, 1978.

ABOUT THE AUTHORS

Genadii Germanovich Bondarenko– *Moscow State Institute of Electronics and Mathematics (Technical University), Doctor of Physics and Mathematics, Professor. An expert in area of radiation solid state physics.*

Sergei Mikhailovich Zhdanov– *Kaluga branch of Moscow State Technical University Candidate of Technical Sciences, Associate Professor. An expert in the area of materials for electronics.*

Aleksei Panteleevich Korzhavyi– *Metallurgical Works "Amethyst" (Kaluga), Doctor of Technical Sciences, Professor. An expert in the area of materials for electronics.*

Alexandr Nikolaevich Tikhonov– *State Research Institute for Informational technologies and telecommunications, doctor of Technical Sciences, Professor. An expert in the area of materials of electronics.*

METHOD FOR CONTROLLED DEPOSITION OF CARBON FILMS

A.M.Baranov

The problems connected with developing technology for carbon film deposition by ion-plasma methods for obtaining the layers substituting SiO_2 were considered. The results of investigation of carbon films during their growing and etching in real time by means of X-ray scatterometry method have been presented. The film thickness, growth rate, density and roughness of the surface were calculated from time dependence of reflection coefficient of X-radiation of wavelength 1.54 Å. The technique of obtaining the carbon films with ultra-smooth surfaces has been proposed. It has been shown that parameters of investigated layers do not change during heating up to 550^0C.

INTRODUCTION

At the present time, in the silicon microcircuits, SiO_2 is used as the dielectric layer, and aluminum is used as contact material. However, during transition from an achieved integration level of 0.25 to 0.18 μm and further (0.13 μm, 0.09 μm), the use of SiO_2 and Al does not lead to the increase of the operating speed because of the presence of transient processes, which depend on the product R×C (resistance multiplies capacity). Therefore many companies started to work on substituting an aluminum with copper [1] (copper resistance is half of aluminum). But this step is only the first stage. It is planned further to substitute SiO_2, ε=3.9-4.3, with materials of dielectric constant less 3 (so-named low-k dielectric). The attractiveness of such a change of technology is that one can simultaneously decrease the response time and halve the number of the levels of metallization. That, in turn, simplifies and reduces the price of microcircuits manufacturing. The carbon films are in the group of the main candidates (different porous materials and composite glass are also in this group), claiming to the place of SiO_2. The carbon material's ε is known to change from 5.7 for diamond to 1.9 for teflon. A lot of carbon materials were studied with the object to use them as the dielectric layers [2]. The films, synthesized in glow high-frequency discharge from gas phase, are of special interest, since they can be obtained as the thin layers on the substrates of large area. Particularly, it has been determined that ε of hydrogenated carbon films (a-C:H) changes from 2.7 to 3.8, and from 2.1 to 2.8 for the fluorine-containing (a-C:H). The obtaining of films with ε<2 is strategic direction. However, the use of carbon films as the isolating layers requires solving the series of important problems, connected with natural peculiarities of carbon films itself.

In contrast to SiO_2, the properties of carbon films can change within wide range, depending on the conditions of the obtaining (Table) [3]. However the carbon films, deposited in plasma were investigated for a long time, the main direction of investigations was to obtain the solid films, which can be used as the shielding coatings to the mechanical effects. The investigation of properties of sub-micron carbon films, in particular the investigation of the beginning stages of their growth, was not considered enough [4].

Table : The turn-down of carbon films properties, obtained by ion-plasmous methods (sputtering, deposition from ion beam and from gas phase in plasma)

#	Parameter	Turn-down
1	Density, g/cm^3	1.2-3.5
2	Specific resistance, Ω·cm	1-1 10^{14}
3	Dielectric constant	2-5.7
4	Optical forbidden gap, eV	0.5-3.5
5	Refraction coefficient in visible range	1.5-2.8
6	Hardness by Vikkers, kg/mm^2	to 6000
7	Internal strength, GPa	to 40
8	Amount of hydrogen, fluorine, %	0-60

The changeability of carbon films properties is the attractive factor, as far allows to form the layers of required properties. But this circumstance complicates significantly their practical use, since a little change of any technological parameter during the growth can lead to the strong change of both the structure and, consequently the properties of film, and the film growth rate. It causes the necessity of changing the traditional approach in controlling the parameters of carbon films and, particularly, the development and use of diagnostic methods, allowing continuously to measure the films parameters during their synthesis (*in-situ* measurements). But none of films parameters measurement method allows to carry out *in-situ* the monitoring of sub-micron carbon films parameters during their growth, especially at the beginning stages of the growth [5,6].

The thermal stability of films is the second main problem. In order to use the carbon coatings as the isolating layers, it is necessary all the parameters should be stable at temperatures more than 400^{0}C, at which the part of production operations is performed during the manufacturing of microcircuits. This temperature is the top limit of stability for the most amorphous carbon films. At a higher temperature the structural changes of the films, which go with the outlet of the hydrogen and fluorine from films, start occurring. As the result, the films are graphitizing and the dielectric permeability and conductivity increases.

The aim of this work is to consider the problems connected with using the new diagnostic methods in the carbon films production, in particular, to use the X-ray scatterometry for monitoring the parameters of carbon films at the beginning stages in real time, as well as to investigate the ways of increasing the temperature stability of carbon films.

EXPERIMENTAL RESULTS

The films containing more than 50% of hydrogen have the minimum dielectric permeability. Therefore exactly these films have the chance to be claimed. The problem is that by increasing the amount of hydrogen and fluorine, one should not reduce the process of their growing to the one of polymerization mechanism in the plasma [7], and reduce the films themselves to the polymer-similar ones, which are of low temperature stability.

Using the hydrocarbons as the working gas with great ratio H/C, one can produce the carbon films with large amount of hydrogen [8]. In a given work, the films a-C:H, obtained by means of magnetron sputtering of graphite target, deposited from gas phase in the plasma excited by high-frequency discharge (13.56 MHz) and from ionic source, have been investigated. The vapors of cyclohexane (C_6H_{12}) and the mixture C_6H_{12}+Ar have been used as the working gas contained the carbon atoms. As it has been shown early, the hydrogen amount of in the similar films is about 50% [9]. The silicon plates of initial average quadratic roughness of 8A were used as the substrates.

Carbon Films Parameters Control in the Process of their Growing

One can control the growth of layers SiO_2 by means of ellipsometry method. However, the practice shows that ellipsometry is bad candidate for measuring the parameters of carbon films due to the indeterminacy of their optical constants, partial light absorption and possibility of creation of an uncontrolled transition layers on the boundaries of carbon films-silicon with an unknown parameters. Therefore for controlling the parameters of sub-micron films during the growth, the method based on the registration of time dependence of reflection coefficients of X-radiation on the wavelength of 1.54 nm was proposed [10,11]. The details of methods of obtaining the films and the construction of X-ray optical system are described in works [5,10].

The interference of X-radiation reflected from the boundaries of vacuum-film and film-substrate, which appears while changing the film thickness is the base of X-ray scatterometry. The technique of *in-situ* X-ray scatterometry is close to laser interferometry in one's own substance. The use of wavelength 1.54 Å allows to study the growth of films starting from 10 Å. Thus, not only the films thickness, but also density and roughness of films, i.e. the parameters which are very importance for sub-micron films, are measured. The films density serves as the measure of quality, since it is connected with the structures of films, with the amount of pores and other inclusions, the presence of which changes the dielectric constant and the resistance. The roughness of surface is the key during deposition of the metallization.

The change of thickness d leads to the periodic oscillation of reflection coefficient (Fig.1). The oscillation-repetition cycle L is determined by Bregg's equality [12]. The change of roughness of film surface leads the change of oscillation amplitude and their average value. Therefore by analyzing the shape of experimental curve I(t), one can calculate the density ρ and surface roughness σ immediately in the process of film growth (for example, by solving the inverse problem and by approximate methods [13]). The main advantage of a given control method is that it allows to visualize the process of forming the film coating with the help of harmonic signal by perfection of which one can judge about the perfection (stability) of the film growth process itself. As the result, one can not only control the parameters of the growing film, but also control the deposition process.

In Fig.2 (curve 1), the experimental dependence of intensity I of reflected beam vs. time of deposition of carbon film I(t) on the silicon plate is shown. The deposition was carried out in high-frequency discharge. It is seen that dependence I(t) oscillates along the film growth. After having finished the process growth, the oscillations I(t) are not observed. The growth rate (V) is defined from formula $V=(L/2)(t_{m+1}-t_m)$, where t_m is time corresponding to peaks m=1,2,.... The film thickness was calculated from formula $d=(L/2)m+V(t_i-t_m)$, where t_i is the running time.

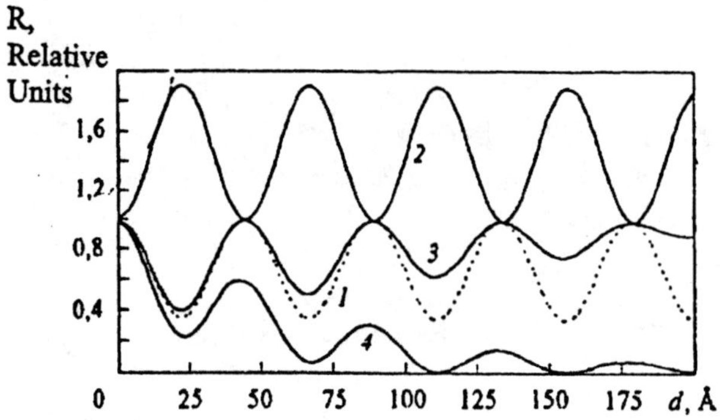

Fig.1 Theoretical dependencies of reflection coefficient R vs. thickness d at the wavelength 1.54 Å for silicon substrate (ρ_1=2.33 g/cm^3 and σ_1=5 Å) and films with parameters: 1 – ρ_2=1.9 g/cm^3; and $\sigma_2=\sigma_1$=5 Å; 2 – ρ_2=2.8 g/cm^3; and $\sigma_2=\sigma_1$=5 Å; 3 – ρ_2=1.9+0.0014d and $\sigma_2=\sigma_1$; 4 – ρ_2=1.9 g/cm^3 and σ_2=5+0.06d.

The dependence of intensity of the beam reflected from system silicon-carbon film during etching of film is shown in the same figure (curve 2). The etching was carried out in the plasma excited by high-frequency voltage. The oxygen was employed as the working gas. It is obvious that the etching curve 2 replicates the curve of deposition curve, but goes upper of the first and changes with much lower rate. The disappearance of oscillations indicates on the complete removal of the film from the substrate surface and allows to define the etching completion moment.

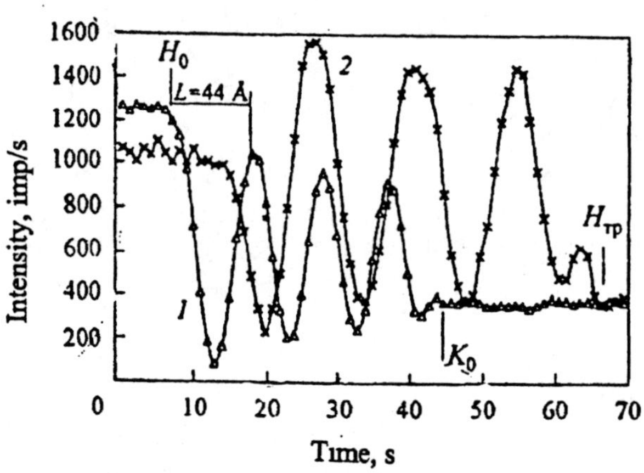

Fig.2 Reflected X-radiation intensity dependencies vs. the time of carrying out the processes of deposition of carbon film (1) and its etching (2) (inverse time scale). H_0 is the beginning of deposition; K_0 is the end of deposition; H_{et} is the beginning of etching.

The change of the oscillations contrast and average value of received dependencies indicate that the density and roughness are not constant during deposition or etching. The calculated dependencies of σ and ρ during deposition and etching of the film are presented in Fig.3a,b [14]. It is seen that while growing the film, the density and roughness increases from 1.53 to 9 1.9 g/cm^3 and from 8 to 9 Å, respectively. When the film thickness is more than 120 Å its density and roughness are practically an unchanged. Vice versa, in the process of the etching with decreasing the film thickness from 66 Å, its density is unchanged. It is important to note that film density is much greater of polymers density (polyethylene, for example).

The films growth rate depends on the pressure of the working gas and gas mixture composition. Therefore the investigations of dependence of carbon films growth rates vs. pressure of cyclohexane and amount of Ar in gas mixture $Ar+C_6H_{12}$ have been carried out. The obtained dependencies are presented in Fig.4. It is seen from Fig.4a that in regions of low pressures, the film growth rate increases from 3.1 to 7.8 Å/s with increasing the pressure of C_6H_{12}. Having achieved the peak value at $P=2.6\cdot10^{-2}$ mm of mercury, the growth rate begins slowly decreasing to 5.5 Å/s when increasing the pressure to $P=5.8\cdot10^{-2}$ mm of mercury. The further increase of pressure does not lead to changing the growth rate. It should be noted that the change of pressure did not practically lead to the change of density and roughness of films. It is obvious that a given result is the sequence of the change of parameters of high-frequency plasma, in particular, the change of concentration of active particles and their mobility. For more detailed explanation of an obtained result, one should investigate the plasma parameters. The behavior of growth rate dependence vs. pressure of the working gas is obviously typical during the deposition of films from gas phase in plasma of high-frequency discharge.

Adding Ar at constant pressure of cyclohexane leads to decreasing the films growth rate. This probably is caused by sputtering the growing carbon film by Ar ions. Two dependencies at partial pressure of cyclohexane of $2.6\cdot10^{-2}$ mm of mercury (Fig.4b, curve 1), at which the growth rate of carbon film is maximal, and at $5.8\cdot10^{-2}$ mm of mercury (Fig.4b, curve 1) have been obtained. It is seen that at cyclohexane pressure of $5.8\cdot10^{-2}$ mm of mercury the growth rate changes slightly when increasing the partial pressure of Ar. When the pressure of cyclohexane is $2.6\cdot10^{-2}$ mm of mercury (Fig 4b, curve 1) the growth rate decreases in a half when decreasing the partial pressure of Ar to $3.6\cdot10^{-2}$ mm of mercury, and then is practically unchanged. It should be noted that growth rate stopped changing at pressure in the camera of more that $6.2\cdot10^{-2}$ mm of mercury, that exactly corresponds to pressure of cyclohexane (Fig.4a) at which the growth rate also stops changing. A much stronger decrease of the rate at cyclohexane pressure of $2.6\cdot10^{-2}$ mm of mercury can be explained by two circumstances: by the negative bias voltage on electrode, on which the substrates are mounted, is doubled (-200V) and by that the percentage of Ar in gas mixture is larger. As the result, the rate of sputtering the growing film by Ar ions is higher. The change of the roughness of film surface at cyclohexane pressure $2.6\cdot10^{-2}$ mm of mercury (Fig.4c, curve 1) and practically the unchanged value of roughness at cyclohexane pressure of $5.8\cdot10^{-2}$ mm of mercury (Fig4c, curve 2) is the confirmation of that Ar ions sputters the growing film. The density of films has not practically changed in all the cases.

The Polishing of the Carbon Films Surface

As it is seen from Fig.3a, during the etching in the oxygen, the roughness of surface of diamond-like film is lesser than the roughness of initial silicon plate (excepting the layers near Si substrate, that is obviously connected with influence of substrate). The decrease of carbon film surface roughness is very important, since this effect can to proposed as a basis of the planing technique of materials surface, which has been already polished by other methods. As the result, one can not only obtain the carbon films with super-smooth surface but also plane the surface Si after its treatment in the plasma, carried out with purpose of the clearing or etching. It is important to note that carbon films are easily taken to polishing in high-frequency plasma in oxygen.

The planing technique is the deposition of carbon film on initial substrate surface of another material, and then the carrying out of the ion-plasmous polishing etching of this film, but not the substrate surface itself [15]. As the result, one obtains the substrate with film, the surface roughness of which is lesser than of substrate itself. Thus the film thickness can be few tens of Angstroem units.

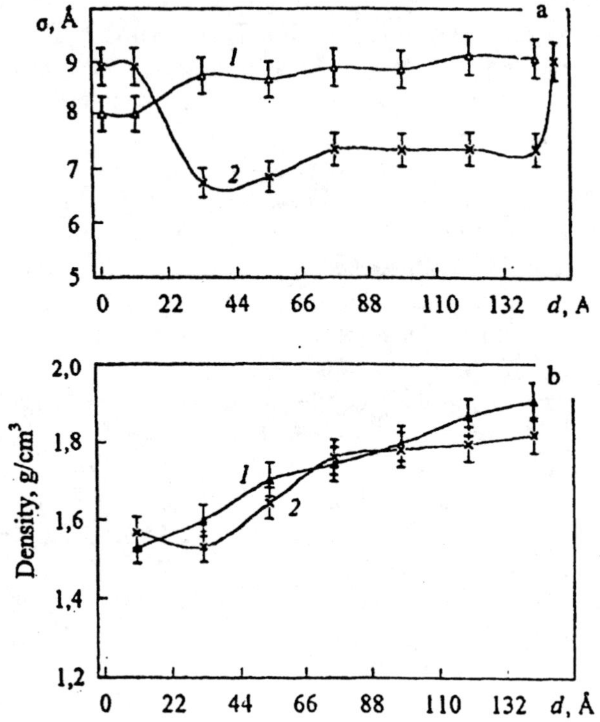

Fig.3 The calculated values of roughness of film surface (a) and its density (b) during deposition (1) and etching (2).

Thermal Stability

The carbon films thermal stability is directly connected with their structure. But it is difficult to register the changes in structure of carbon films during the heating, since the films are amorphous. One can go by other way – to investigate not single-layer films, but multi-layer carbon periodic structures (MS) with the layers of nanometer thickness. The peaks, which are similar to Bregg's peaks of crystalline materials are observed on angle dependence of X-radiation reflection coefficient. Due to the change of intensity, half-width and Bregg's peak placement, one can judge about stability of layers MS and boundaries between monolayers during the heating.

Multi-layer carbon structures were obtained by means of sequence deposition of layer by megnetron sputtering method and ion-plasmous deposition from mixture $Ar+C_6H_{12}$ in different ratios of Ar and C_6H_{12}. The control of the layers growth during the formation of multi-layer structure was accomplished on the base of data obtained by means of *in-situ* X-ray scatterometry method [16].

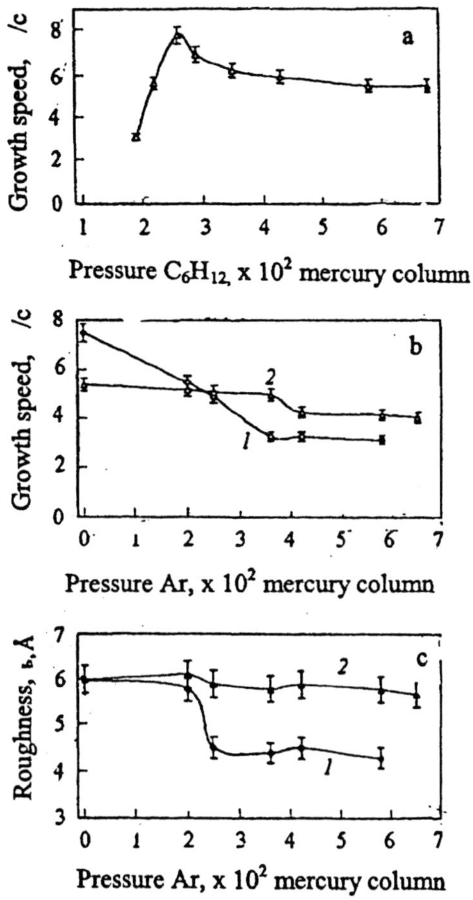

Fig.4 The dependence of growth rate of carbon films vs. pressure of C_6H_{12} (a), dependencies of growth rate (b) and roughness of film surface (c) vs. pressure of Ar in mixture C_6H_{12}+Ar. The pressure C_6H_{12}=2.6·10^{-2} mm of mercury (1) and 5.8·10^{-2} mm of mercury (2).

The study of thermal stability of MS was carried out by means of their annealing in vacuum [17]. MS was kept at a given temperature for one hour. After every stage of treatment the measurement of angle dependence of reflection coefficient was carried out. The temperature dependencies of parameters of multi-layer structure with period 94 Å and period number of 30 are presented in Fig.5. It is seen that parameters of MS remain practically constant even at 550^{0}C, and consequently, the parameters of individual layers engaged in MS are unchanged. That is much greater of a known results received early [2]. Stability of artificially created periodic carbon structure with monolayers of nanometer thickness is obviously higher than of homogenous carbon films. The change of parameters of MS in temperature range from 200 to 300^{0}C is connected with the processes of the rebuilding of amorphous structure of carbon films. In fact, any films, obtained by deposition from plasma at room temperatures, are growing under nonequilibrium conditions and usually are in need of the annealing which stabilize their parameters. Within range from 300 to 550^{0}C, the parameters of MS are unchanged. The temperatures higher 600^{0}C were not employed in a given investigation, since MS was formed on glass substrates, the softening temperature of which is in the region 600^{0}C. He annealing of MS on the air at temperature higher 300^{0}C led to their damaging (reflection coefficient sharply decreases). That is the consequence of damaging influence of oxygen on carbon layers.

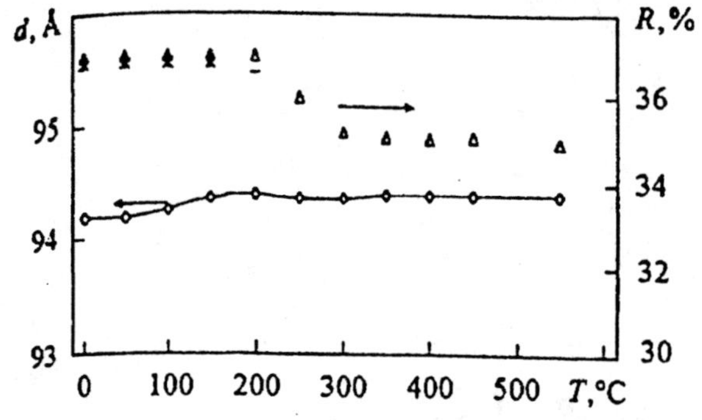

Fig.5 Dependence of period d and reflection coefficient R vs. etching temperature for MS.

CONCLUSIONS

1. X-ray scattery technique has been used for investigating the parameters of carbon films in the process of their formation. The film thickness, growth rate, density and roughness of the surface were calculated from time dependence of reflection coefficient of X-radiation on the wave length of 1.54 Å.
2. It has been shown that density and roughness of carbon films can be strongly changed not only dependently on production conditions, but also at the beginning stages of growth.
3. It has been shown that during etching in the oxygen, the surface roughness of carbon films can be lesser than roughness of initial film. At the result, one can obtain the substrates with supersmooth surface.
4. It has been determined that parameters of the investigated carbon layers do not change at heating up to 550^{0}C.

REFERENCES

1. Singer P. Semiconductor International, 1998, no.6, p.91-98.
2. Baliga J. Semiconductor International, 1998, no.6, p.139-144.
3. Tsai H., Body D.B. J.Vac.Sci.Technol. A, 1987, v.5, p.3287-3312.
4. Lifshitz Y. Diamond and Related Materials, 1996, v.5, p.388-400.
5. Baranov A.M., Sleptsov V.V., Tereshin S.A., et al. Proc. SPIE,1995,v.2519,p.108-115,
6. Baranov A. M., Tereshin S.A., Mikhailov I.F.et al., Proc. SPIE, 1996, v.2863, p.359-367.
7. Yasuda H. Plasma Polymerisation, Academic Press, Inc., New York, 1985, p.255.
8. Grishko L.B., Elinson V.M, Sleptsov V.V., i dr'. Elektronnaya promyshlennost, 1986, no.4, s.6-11.
9. Kondrashov P.E., Smirnov I.S., Baranov A.M., i dr. Trudy Vserossiyskoy konferentsii"Almazy v tehnike i elektronike", Moscow, Polyaron, 1997, s. 89-90.

10. Mikhailov I.R, Pinegin V.I., Sleptsov V.V., et al. Crystal Research and Technology, 1995, v.30, no.5, p.643-649.
11. Baranov A.M., Kondrashov P.E., Smirnov I.S. Trudy Mezhdunarodnoy konferentsii "Almazy v tehnike i elektronike", Moscow, Polyaron, 1998, s. 141-149.
12. Mikhailov I.F., Borisova S.S., Fomina L.P., et al. Proc. SPIE, 1995, v.2453, p.186-193.
13. Baranov A.M., Mikhailov I.F. Thin Solid Films 1998, v.324,no.l-2,p.63-67.
14. Kondrashov P.E., Smirnov I.S., Baranov A.M., et al. Diamond and Related Materials, 1997, v.6, p.1784-1788.
15. Baranov A.M., Surface and Coating Technology, 1998, v.l02,no.l-2,p.l54-158.
16. Kondrashov P.E., Smirnov I.S., Novoselova E.G., et al. Diamond and Related Materials, 1997, v.6, p.902-905.
17. Kondrashov P.E., Smirnov I.S., Baranov A.M., etal. Diamond and Related Materials, 1998, v.7, p.1647-1650.

ABOUT THE AUTHOR

Baranov Aleksandr Mikhailovich *– State Research Institute of Vacuum Technology after S.A.Vekhshinkii. Candidate of Physics and Mathematics, team manager. An expert in area of technology of diamond-like carbon films.*

STOICHIOMETRICAL COMPOSITIONS AND CRYSTAL-CHEMISTRY REGULARITIES OF SELENITE FORMATION

N.G. Goraschenko, S.Yu. Tsareva, Zh.S.Kuchuk and A.A.Mayer

A model of the crystal-chemistry approach for the selenites class is presented. This model allows evaluating the possibility for formation of selenite with the cross-linking ions of different valence and to calculate the maximum possible boundaries of the homogeneity area. The application of this model to Bi_2O_3-MgO and Bi_2O_3-CuO systems has been demonstrated. The parts for equilibrium diagrams of these systems are built.

INTRODUCTION

Crystals with the structure of selenite have important physical features. Due to electro-optic, magneto-optic, piezoelectric and acoustic properties, wide range of transparency, they are used as the continuous and impulse electro-optic modulators of light, polarizations switches in electro-optic deflectors of light beams, in light-valve devices (space-time light modulators) and as the reversing working mediums of optical memory devices of high capacity and in other devices of functional electronics. The possibility of controlling the physical properties of selenites by means of doping the cross-linking ions (which stabilize the metastable body-centered cubic structure γ-Bl_2O_3) and the alloying ions makes the selenites a prospect material which does not loose the actuality. However, in the practice, only silicon-, germanium-, titanoselinites and solid solutions on their base have found the practical use. Meanwhile the significant number of compositions and solid solutions can be crystallized in structural type of selenites that extends the possibility of researchers. Probably it is explained by the absence of single-valued opinion about composites of selenites and their solid solutions, the absence of data on the equilibrium diagrams of selenites of different systems with the boundaries of homogenity. Therefore the proposed crystal-chemistry approach can extend the spectrum of investigated selenites systems for practical purposes, since it makes significantly easy the work on determining the possibility of formation of compounds and solid solutions of selenite structure and their existence range during the plotting of the equilibrium diagrams.

The model of structure, proposed by Craig and Stepherson [1] for Fe- and Zn-selenites, is the basis crystal-chemical approach: 24 bismuth-oxygen disfigured polyhedrons form the "framework" $Bi_{24}O_{40}$, containing two tetrahedron voids, which are infilled with ions of elements (further they will be named as the cross-linking elements). The ions Bi^{3+} and Bi^{5+} are the cross-linking elements for γ-Bi_2O_3; ions Si^{4+} or Ge^{4+} are cross-linking elements for Si or Ge-selenites, ions E^{n+} (2 n 5, n 4) and Bi^{3+} or Bi^{5+} are the cross-linking elements in selinites with cross-linking elements the valence of which differs from 4. This model allows to consider the formation of compound with the structure of selenite, resulting from infillment of tetrahedron voids by ions of the cross-linking elements of valence from 2 to 5; The metastable modification γ-Bi_2O_3 is the limit structure of selenite, in which the ions Bi^{3+} and Bi^{5+} are placed in tetrahedron voids.

One should consider the fact of presence of ions Bi^{5+} in structure γ–Bi_2O_3 or in structures of some selenites (for example, zinc). Some authors state [2] that they did not succeed in ascertaining the fact of ions Bi^{5+} presence in structure of selenite, by using chemical analysis method; other authors [3] explain the presence of certain properties (for example, the thermostimulated luminescence) by the presence of ions Bi^{5+} in structure. Thus, the question of presence of ions Bi^{5+} in selenite structure is open. The consideration of γ-Bi_2O_3 and some selenites as the framework $Bi_{24}O_{40}$ with ions Bi^{5+} and the cross-linking tetrahedron voids in our model of crystal-chemical approach gives a real possibility of predicting the stochiometrical composition and area boundaries of selenites homogenity, that will be demonstrated further for concrete examples.

The Model of Crystal-Chemical Approach for Selenites Class

During the infilling of tetrahedron voids by ions of the cross-linking elements of valence 4, the ideal structure of selenite with thoroughly infilled cation and anion positions is formed. During the infilling of tetrahedron voids by ions of the cross-linking elements of valence not 4 or during partial infilling of tetrahedron voids by ions, the structure must contain the partial fulfilled anion positions or compensate the charge in other way (for example, the ions Bi^{3+} or Bi^{5+} in tetrahedron positions). The similar consideration of selenite structure allowed to write the quasichemical equations of possible mechanisms of formation of selenites with cross-linking ions of different valence and to determine the possible theoretical boundaries of selenites homogenity areas [4]. As the example, let us write quasi-chemical equation of stoichiometrical composition selenite formation with bivalent cross-linking ions:

$$Bi^{3+}+Bi^{5+} \quad 2E^{2+}+V_O^{2-}.$$

The composition with tetrahedron voids infilled thoroughly by ions of the cross-linking elements is considered to be of stoichiometrical composition. The homogenity area is on the side of the stoichiometrical composition and can be represented as the demonstration of γ-unstoichiometry (unbalance between cation crystal components (Bi^{3+} or Bi^{5+}) and E^{n+} (2 n 5)) and σ-unstoichiometry (deficiency or excess of oxygen) [5].

The stoichiometrical compositions and theoretical boundaries of homogenity areas of selenites, depending on the valence of the cross-linking ion are presented in Table 1.

It should be noted that a given evaluation of the length does not take into consideration the nature of ions, and in the real systems the area boundaries can be more narrow, than these given in Tale 1. However, the tendency to the expansion of boundaries of selenite homogenity while deviating the valence of cross-linking ion from 4, maintains.

Table 1 Stoichiomerical compositions and theoretical boundaries of homogenity areas of selenites depending on valence of the cross-linking ion

E^{n+}	Stoichio-metrical composition	Homogenity boundaries areas	Length of homogenity area, mol.%
E^{2+}	$Bi_{12}EO_{19}$	6:1-25:3	10.7
E^{3+}	$Bi_{24}E_2O_{39}$	12:1-24:1	3.9
E^{4+}	$Bi_{12}EO_{20}$	6:1-24.5:3	3.4
E^{5+}	$Bi_{24}E_2O_{41}$	12:1-25.5:0.5	5.8

Further, the lattice parameter differs in the selenites with different cross-linking ions and within area of selenite homogenity. The linear variation of the lattice parameter within area of homogenity, probably, can be explained by the change of the kind and number of ions (Bi and the cross-linking ions), occupying the tetrahedron positions in structure.

In first approach, one can suppose that lattice parameter depends on radius of the cross-linking ion. Based on an above-stated, one should be logically to assume that character of variation of seletites lattice parameter with different cross-linking ions is linear, when take into account the discordance of ion valence and its coordination number in structure.

In work [4], such a dependence of lattice parameter vs. reduced ion radius (which is calculated as r·S', r is ion radius taken from tables [Å]) with taking into account the amendment S' has been proposed. S' is some averaged sum of valent forces of cations in structure for 1 anion:

$$S' = \frac{\sum_i V_{kat} / Q_{kat}}{\sum atoms}$$

where V_{kat} is valence and Q_{kat} is coordination number in structure of cation.

Table 2 Values of the amendments S' for the calculation of reduced radii of the cross-linking ions of the different valence

E^{n+}	Stoichiometrical composition	S'
E^{2+}	$Bi_{12}EO_{19}$	0.297
E^{3+}	$Bi_{24}E_2O_{39}$	0.302
E^{4+}	$Bi_{12}EO_{20}$	0.307
E^{5+}	$Bi_{24}E_2O_{41}$	0.312

The values of amendments S' for stoichiometrical compositions of selenites of different valence of the cross-linking ions are given in Table 2.

In the Fig.1, the dependence a=f(rS') is shown. This dependence is built in the following way:

1. The lines for the stoichiometrical compositions of selenites with three- and four-valent cross-linking ions were drawn through dots (rS'; a) for $Bi_{24}Al_2O_{39}$, $Bi_{24}Ga_2O_{39}$, $Bi_{24}Co_2O_{39}$, $Bi_{24}Fe_2O_{39}$ and $Bi_{12}SiO_{20}$, $Bi_{12}GeO_{20}$, $Bi_{12}TiO_{20}$ and γ-Bi_2O_3, the values a=10.268 Å and rS'=0.261 Å of which are the limit for stable selenites.

Besides, for the dependence a=f(rS') it is possible to draw the lines for stoichiometrical compositions of selenites with bivalent and quinquivalent cross-linking ions at the presence of necessary amount of data about lattice parameters of these selenites.

2. The boundaries of metastable area are determined by lattice parameters a=10.268 Å and by value of reduced ion radius rS'=0.261 Å for γ-Bi_2O_3.
3. For the dependence of lattice parameter vs. an reduced ionic radius (Fig.1), the following areas are marked out:

   I. the area of existence of stable selenites (a<10.268 Å, rS'<0.261 Å).
   II. the area of existence of metastable selenites (a>10.268 Å, rS'<0.261 Å).
   III. the area , where the selenites formation is impossible (rS'>0.261 Å).

On the base of an above-said, the aspects of crystal-chemical approach being supported on which, one can forecast the possibility of formation of stoichiometrical compositions selenites, can be formulated in the following way:

1. The structure of selenite can be represented as the framework $Bi_{24}O_{40}$ formed by the distorted bismuth-oxygen polyhedrons, and which contains two tetrahedron voids.
2. During the infilling of these tetrahedron voids by ions Bi^{3+} and Bi^{5+}, the limit structure of selenite with lattice parameter a=10.268 Å is formed.
3. During the infilling of the voids by ions of quadrivalent elements (Si,Ge), the ideal structure of selenite with the thoroughly fulfilled anion and cation positions is formed.
4. For the partial infilling of tetrahedron voids or for their infilling by ions of the cross-linking elements of valence not 4, the structure must either contain the partial infilled anion positions, or compensate the charge by other way (by ions Bi^{3+} or Bi^{5+} in tetrahedron positions, for example).
5. The homogenity area can be represented as behavior of γ-unstoichiometry (unbalance in the ratio between cation components of the crystal (ions Bi (Bi^{3+} or Bi^{5+})) and σ-unstoichiometry (deficiency or excess of oxygen).
6. A linear behavior of change of lattice parameter within homogenity area is explained by the change of kind and ions number (of Bi and cross-linking ions) in tetrahedron positions and is the function of size of cross-linking ion in first approach.
7. The dependence (Fig.1) of selenites lattice parameter vs. reduced ion radius with taking into account the amendment S' (Table 2) is linear.

   The boundaries of stable selenites formation is determined by value of lattice parameter a=10.268 Å and the reduced ion radius rS'=0.261 Å for γ-Bi_2O_3.

   If the value rS' for a given ion exceeds value 0.261 Å, than the selenite structure formation with this ion being the cross-linking ion is impossible.
8. Writing the quasi-chemical equations of a possible mechanisms of selenites formation and their homogenity areas (as γ-unstoichiometry). The possible boundaries of selenites homogenity areas, depending on valence of the cross-linking ion has been theoretically revealed (Table 1).

   Basing on the above-said states, the crystal-chemical approach allows to evaluate the possibility of selenite formation with the cross-linking ion due to dependence a=f(rS'), to determine its stoichiometrical composition and the approximate boundaries of homogenity area (Table 1).

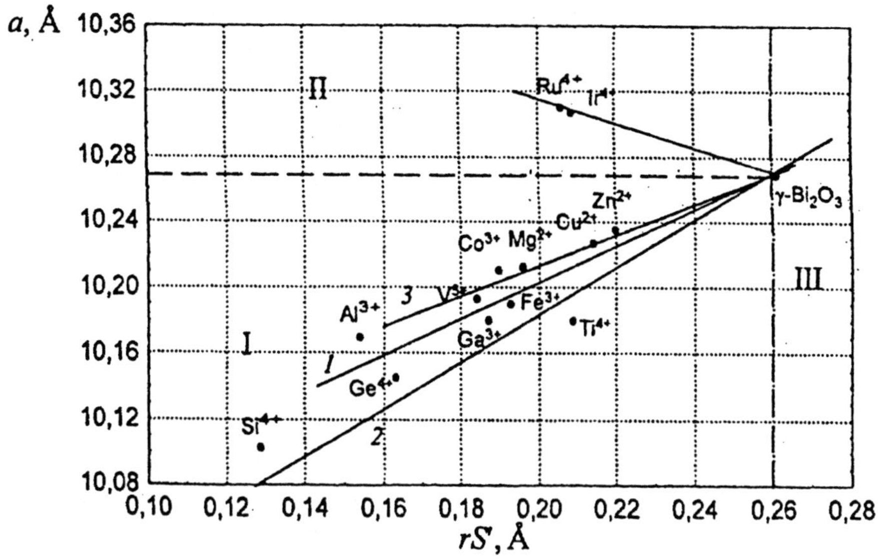

Figure 1 The dependence of selenites lattice parameter vs. the reduced ion radius with the plotted dependencies a=f(rS') for selenites with bivalent, trivalent and quadrivalent cross-linking ions: I – existence area of stable selenites (a<10.268 Å; rS'<0.261 Å); II – existence area of metastable selenites (a<10.268 Å; rS'<0.261 Å), III – the area where the selenite structure formation is impossible (rS'>0.261 Å). 1 is the line for $Bi_{24}E_2^{3+}O_3$; 2 is the line for $Bi_{12}E^{4+}O_{20}$, 3 is the line for $Bi_{12}E^{2+}O_{19}$.

Crystal-Chemical Forecasting of Selenite Formation Possibility with Bivalent Cross-Linking Ion

A some data about selenites with bivalent cross-linking ions are given in literature [6-9], but there are not the single-valued opinion about possibility of selenites formation in these systems.

The desire to make this question clear and to obtain the data for plotting the dependence a=f(rS') for selenites with bivalent cross-linking ions has determined the subjects of inquiry, being the selenites in systems Bi_2O_3-MgO, Bi_2O_3-CuO, Bi_2O_3-CaO, Bi_2O_3-BaO, Bi_2O_3-CdO.

In order to evaluate the possibility of selenites formation in these systems it is necessary to calculate the value of the reduced ion radii for Mg^{2+}, Cu^{2+}, Ca^{2+}, Ba^{2+}, Cd^{2+}. The results of calculation are presented in Table 3, which shows that values of the reduced radii for Mg^{2+} and Cu^{2+} are less of maximum permissible value for the selenites existence, and exceeds this value for Ca^{2+}, Ba^{2+}, Cd^{2+}.

Table 3 The stoichiometrical compositions and the calculated values of the reduced ion radii for ions of some bivalent elements

E^{2+}	Stoichiometrical composition	*S*	*r*, Å[10]	*rS*, Å
Mg^{2+}	$Bi_{12}MgO_{19}$	0.297	0.66	0.196
Cu^{2+}	$Bi_{12}CuO_{19}$	0.297	0.72	0.214
Ca^{2+}	$Bi_{12}CaO_{19}$	0.297	0.99	0.294
Ba^{2+}	$Bi_{12}BaO_{19}$	0.297	1.34	0.398
Cd^{2+}	$Bi_{12}CdO_{19}$	0.297	0.97	0.288

Theoretically calculated values of homogenity area boundaries in systems Bi_2O_3-MgO and Bi_2O_3-CuO must extent from 3.8 to 14.2 mol.% MgO, CuO, respectively.

The selenites have been obtained by means of solid-phase synthesis by heating the oxides mixtures Bi_2O_3 and MgO, Bi_2O_3 and CuO at t=760^0 and 750^{0}C, respectively, during ~10-30 hours. The completeness of reaction was checked by X-ray phase analysis (XPA).

The lattice parameters of synthesized selenites have been calculated. For stoichiometrical compositions $Bi_{12}MgO_{19}$, $Bi_{12}CuO_{19}$, $Bi_{12}ZnO_{19}$ (our data) the line has been drawn over dots (rS',a). This line as well as the line for stoichiometrical compositions with trivalent and quadrivalent cross-linking ions are passing through the dot (0.261 Å; 10.268 Å) γ–Bi_2O_3 (Fig.1) that confirms the appropriateness of a proposed crystal-chemistry model, in that) γ–Bi_2O_3 is assumed to be the limit selenite.

Due to the results of X-ray phase analysis, visual-polythermal, differential-thermal analyses the parts of equilibrium diagrams of systems Bi_2O_3-MgO and Bi_2O_3-CuO (Fig.2,3) have been plotted. Homogenity area in system Bi_2O_3-MgO expanses from 3.8 to 14.2 mol.% MgO, and from 7.7 to 14.2 mol.% Cuo in system Bi_2O_3.

The fact that homogenity area boundaries appears to be a more narrow than a calculated theoretically (Table 1) can be explained by influence of similarity factor of nature of chemical bonding of the interchanging ions Bi^{3+} and Cu^{2+} (γ-unstoichiometry). The electronegativity of ions is used as the similarity factor [11]. The description of influence of isomorphism factors on the solid solutions formation is considered in details in our next article "Use of crystal-chemical approach for studying solid solutions of selenite structure". In Table 4, the values of electronegativity of ions Bi^{3+}, Mg^{2+}, Cu^{2+}, occupying the tetrahedron positions in selenite structure are given. The difference $\Delta\chi$ of ion pair Bi^{3+}/Cu^{2+} is twice much of $\Delta\chi$ for pair Bi^{3+}/Mg^{2+}.

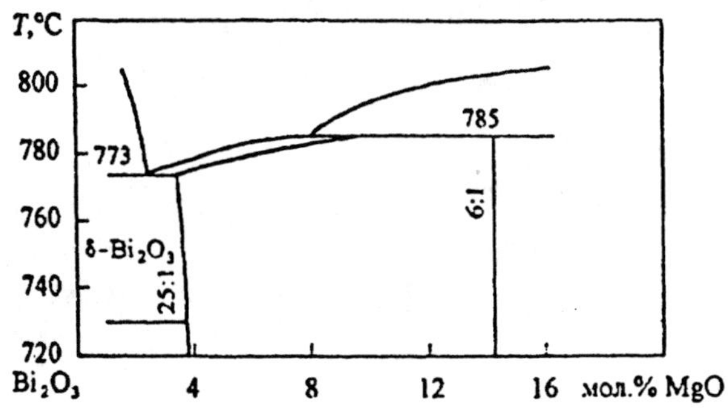

Figure 2 The part of the equilibrium diagram of system Bi_2O_3-MgO.

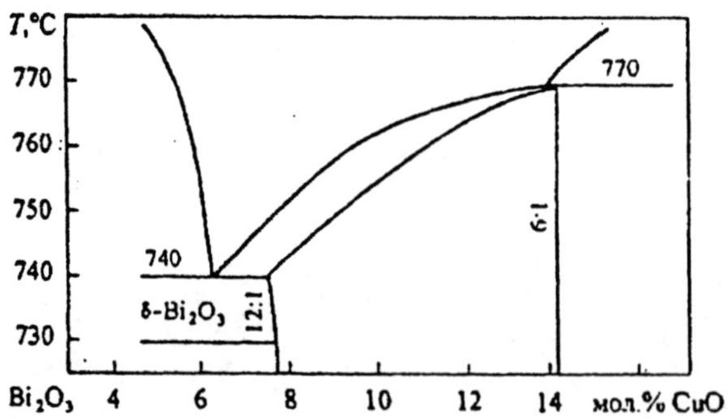

Figure 3 The part of the equilibrium diagram of system Bi_2O_3-CuO.

Table 4 The values of the electonegativities of ions Bi^{3+} and Mg^{2+}, Bi^{3+} and Cu^{2+}, interchanging within homogenity area

Ion pair	Electronegativity χ,kJ/mole [12]	$\Delta\chi$, %
Mg^{2+}	733	
Bi^{3+}	816	10
Cu^{2+}	984	
Bi^{3+}	816	21

Generally speaking, the crystal-chemical approach, taking into consideration only size factor of the cross-linking ions, does not allow precisely to predict the homogenity area boundaries in a given concrete system, but appoints only the boundaries of the possible substitutions E^{n+} (2 n 5) on Bi^{3+} or Bi^{5+}. Under effect of another factors (for example, ions chemical bonding factor) these boundaries can be narrowed, that is what occurred in system Bi_2O_3-CuO. However, the possibility of predicting the existence of selenite structure with any cross-linking ion, having only value of its radius, and the decrease of the number of necessary experiments during the plotting of equilibrium diagrams are the doubtless advantages of such approach, since the maximum possible theoretical boundaries of homogenity area of selenite with the cross-linking ions of different valence are known a priori (Table 1).

Besides, when representing the selenites and their homogenity areas as the result of isomorphic substitution of Bi ions and the cross-linking ions, then all the aspects for simple selenites can be propagated on the systems of solid solutions with the structure of selenite of two and more kinds of cross-linking ions in tetrahedron positions.

REFERENCES

1. Craig D.C., Stephenson N.C. J. of Solid State Chemistry, 1975,v.15,no.1,p.1-8.
2. Kargin Yu.F. Sintez, stroenie i svoystva oksidnyh soedineniy vismuta so strukturoy eillenita. Avtoreferat dissertatsii na soiskanie uchenoy stepeni d. h. n., Moskva, 1998.
3. Denable S., Martinaud M., Schroerer M., Marsaud F., Launay J.C., Hagenmuller P. J. Phys. Chem. Solids, 1990, v.51,no.2,p.171-176.
4. Kuchuk Zh.S. Sintez i svoystva kristallov i plenok sillenitov, soderzhaschih iony perehodnyh elementov Cr^{34+}, Fe^{34+}, Co^{3+}, Ni^{2+}. Dissertatsiya na soiskanie uchenoy stepeni k.h.n., Moskva, 1991.
5. Shimanskiy A.F. Razuporyadochenie i ionnye protsessy v kristallah nestehiometricheskih slozhnyh oksidov. Avtoreferat dokt. diss. na soiskanie uchenoy stepeni d.h.n., Krasnoyarsk, 1998.
6. Kosov A.V., Kutvitskiy V.A., Skorikov V.M., Ustalova O.N., Koryagina T.I. Zhurnal neorganicheskoy himii, 1971, t. 16, no.3,s.863-865.
7. Kargin Yu.F. Zhurnal neorganicheskoy himii, 1994, t.39, no.12, s. 2079-2081.
8. Ramahan A., Gopalakrischnan J., Rao C.N. Mat. Res. Bull., 1981, v.16, no.2, p.169-174.
9. Kargin Yu.F., Skorikov V.M. Zhurnal neorganicheskoy himii, 1984, t.34, no.10, s.2713-2715.
10. Ahrens L. Geochim. Et Cosmochim acta, 1953, part 2, v.2, no.1, p.183.
11. Urusov V.S. Teoreticheskaya krystallokhimiya. Izd. MGU, 1987, s.256.
12. Makarov E.S. Izomorfizm atomov v kristallakh. Atomizdat, Moscow, 1973, s.93.

About the Authors

Natalia Grigorievna Goraschenko– *RCTU. Candidate of Chemistry, Associate Professor. An expert in the area of physical chemistry of solid state.*

Svetlana Yurievna Tsareva– *RCTU. An expert in the area of physical chemistry of solid state.*

Zhanna Semenovna Kuchuk– *Moscow State Technical University (Moscow Engineering Physical Institute). Candidate of Chemistry, Associate Professor. An expert in the area of physical chemistry of solid state.*

Aleksandr Artemievich Mayer– *RCTU. Doctor of Chemistry, Professor. An expert in the area of physical chemistry of solid state.*

CHROMIUM ALLOY WELDABILITY

V.Ya.Panchenko, A.M.Zabelin, A.I.Dedyurin, I.V.Borovitskaya and V.V.Sidorova

The possibility of using a laser for welding and cutting chromium alloys as one of possible candidates for constructional low-activated materials for the units of nuclear fusion reactor has been firstly investigated. The studies have shown that the use of a laser for welding and cutting the chromium alloys is the most prospective, allowing obtaining the necessary width of the weld with a minimal heat-affected area.

At the present time chromium alloys are considered as one of possible candidates for constructional materials for thermonuclear reactors (TNR) [1]. It was mainly connected with their physico-mechanical properties, especially with activation characteristics. According to the results of calculation of the induced activity decay after 10 year of exploration as the first wall of TNR and blanket, the decay time of chromium alloys to the secure level is much smaller than decay time of alloys on the base of vanadium (systems V-Ti-Cr) [1,2].

When working up the chromium alloys for units of TNR the improvement of their technological and mechanical properties (durability, plasticity) is of the greatest interest [1,3,4,5]. Thus the weldability of these alloys which can determine the prospect of their use as TNR materials are not practically considered.

At the present time the argon-arc technique is mainly used for joining the refractory metals, providing reliable enough welded joints with low amount of interstitial impurities [6]. At such a welding technique the heat-affected area great enough is formed, which can reach 500 mm to the width for the welding joint of 4 mm. This requires the reliable protection of this area from its possible pollution by interstitial impurities.

The concentration of heat energy of laser beam is very high, providing heat-affected area local enough [7]. Therefore the use of laser is of doubtless interest. Besides, the change of laser beam diameter allows to control the width of welding joint and, respectively, to decrease the heat affection on near-joint area. Based on mentioned above peculiarities, the use of laser for chromium alloys as the materials sensitive to the heat treatment leading to the significant growth of grains is of great interest. The heat-affected area is narrow enough, that allows to provide a more reliable protection of this area by inert atmosphere (argon, helium) and, respectively, to decrease the effect of interstitial impurity, particularly nitrogen as the most soluble in chrome (for example, 2.8 atom.% of nitrogen and only $4.3 \cdot 10^{-2}$ atom.% of oxygen are dissolved in a chrome at 1500^0C [8]). It should be noted that at present time there are data of using basically the laser for welding the refractory materials, such as Ns, Ta and other, but there are not data on investigating the weldability of chromium alloys [7].

The investigation of the possibility of the cutting and weldability of the plate sample of chromium alloy containing 33.9 atom.%V, 1 atom.%Ti and 0.1 atom%Y by using laser has been carried in a given work. The alloy of a given composition was chosen from the series of the system Cr-V, additionally alloyed with Ti and Y, being of a heightened heat-resistance and of technological effectiveness enough at hot and warm deformation. The additional alloying of the alloys by reactive elements (Ti and Y) was carried out with aim to bind the interstitial impurities such as oxygen and nitrogen, which significantly effect on chromium alloys properties [1,3-5].

The alloy of a given composition was smelted in arc furnace with unspent tungsten electrode on the copper water-cooled bottom at atmosphere of chemically pure argon. The ingot alloy, welded into stainless steel jacket, has been forged to the thickness of 4-4.5 mm at 1100-1200^{0}C with intermediate heating. After proper rounding off, etching and annealing at 1000^{0}C during 1 hour the forging was rolled at temperature 700^{0}C to the thickness of ~1 mm. After warm rolling, the plate was mechanically cleared from oxidized layer and then was etched with 10% solution of oxalic acid. After mentioned above treatment, the investigation of the cutting and weldability of the obtained plates was carried out.

To carry out these investigations the continuous action CO_2 laser TL5M of power 4 kW has been used. The cutting and the welding of plate sample was carried out in the helium medium by means of laser beam of diameter 0.4 mm at the different power and beam movement velocity.

The investigations of cutting the plate have shown that laser beam of power 1.5 kW and velocity of 20 mm/s gives a plane enough cutting surface free of cracks on the edge (Fig.1. The laser welding of power of 3 kW and velocity of 3 m/min has been carried out on one part of the plate rigidly mounted for side faces. The investigation of the appearance of the joint weld has shown that during laser welding the plane junction of the thickness about 0.7 mm is formed, in some parts of which the hot cracks have been revealed. These cracks are spreading to the edge of the plate (Fig.1).

a

Fig.1 Appearance of weld joint surface and the cutting part (a), obtained on the plate sample of alloy Cr-33.9 atom.% V-1atom.%Ti-0.1atom.%Y using gas laser. The below part of weld joint is the cross cracks. ×4.

It is seen from data of microstructure investigations of the weld that junction has cast columnar structure with distinguished accretion area of crystals, forming straight boundary line in the junction center (Fig.2). The presence of accretion area of columnar structure crystals is the one of the imperfections of the weld joint, which can lead to the damage along jointing line. Such a character of junction structure is the evidence of the insufficient looping courses of the regimes of the welding, i.e. not optimal distribution of heat flows.

In near-junction area of length about of 0.18 mm the alloy recrystallization occurred with forming the characteristic polyhedron grains (Fig.2). Outside near-junction area the alloy structure did not practically change and has the elongated grains created during warm rolling. It should be also noted that in the area of weld joint the formation of porosity is not observed as that happens while welding the refractory alloys [9].

The measurement of the microhardness to the section of the weld joint has shown a significant its increase in the central part up to 720 kg/mm^2 in comparison to the initial values of the plate itself – 420 kg/mm^2 (Fig.3). One can note that in the region of recrystallized area a little increase of the microhardness is observed with respect to initial one.

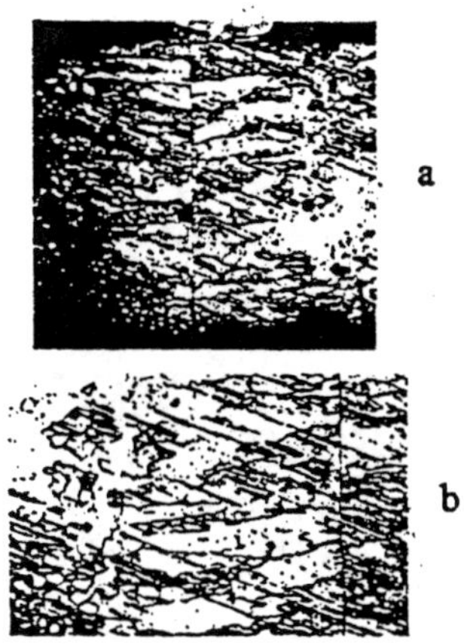

Fig. 2 The microstructure of weld joint with the near-junction area of alloy Cr-33.9atom.%V-1atom.%Ti-0.1atom.%Y. a – ×70; b – ×160.

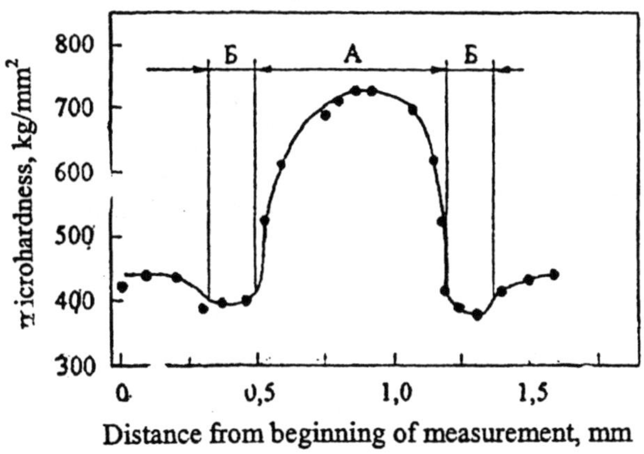

Fig.3 The change of microstructure along section of weld joint and near-junction area of alloy Cr-33.9atom.%V-1atom.%Ti-0.1atom.%Y. A is the weld joint; B is the near-junction area.

A high enough microhardness of plate is caused by its cold-hardening after warm rolling, whereas after annealing at 1300^{0}C during 1 hour it was ~300 kg/mm^2. The significant increase of microhardness in the center of weld joint is probably caused by not so much thermal tensions in the cast structure of the junction as basically caused by its pollution by interstitial impurities that requires additional investigations.

A carried out investigations have shown that high-temperature chromium alloys are satisfactorily treated by laser, regarding both the cutting and the weldability. That gives the reason to suppose that chromium alloys can be welded without significant change of weld joint properties. Moreover during laser welding of the investigated alloy, as seen from a presented data, quite a narrow heat-affected are is formed (~1 mm).

Based on the received results, one can make the following conclusion:

1. For welding the chromium alloys the use of laser is the most prospect, allowing to obtain the necessary width of weld joint with minimal heat-affected area.
2. The formation of optimal structure of weld joint (with interpenetrative elongated grains) and minimal change of its properties can be provided by means of matching the welding regime and protection technique from satiation by interstitial impurities.
3. One should to optimize the chromium alloy composition to complex of properties, which mostly meets requirements demanded for materials of thermonuclear reactor (the hardness, the plasticity, the technological properties and other).

REFERENCES

1. Tsepelev A.B., Bystroe L.N., Savvateev N.N. Perspektivnye materialy, 1995, №3, s.44-47.
2. Ivanov L.I., Platov Yu.M., Dediurin A.I., BorovitskaiaI.V., Lazorenko V.M., Savvateev N.N., Sidorova V.V., Zakharova M.I., Artemov N.A. Journal of Nuclear Materials, 1996, 233-237, p.396.
3. Zakharov M.V., Zakharov A. M. Zharoprochnye splavy. M.: Metallurgiya, 1972, s.308.322.
4. Rakitskiy A.N., Trefilov V.I. Poroshkovaya metallurgiya, 1977, №9, s.62-71.
5. Avtonomov GA., IgolkinaL.S., Kosachev L.S., Mochalov B.A., Rakitskiy A.N., Firstov S.A., Shtauberg M.I. FMM, 1982,t.53,vyp.4,s.814-819.
6. Svarka i svarivaemye materialy. Spravochnik v 3-kh tomakh (pod obschey redaktsiey V.N.Volchenko). T. 1 Svarivaemost materialov (pod redaktsiey E.L.Makarova). M. Metallurgiya, 1991, 528 s.
7. Svarka i svarivaemye materialy. Spravochnik v 3-kh tomakh (pod obschey redaktsiey V.N.Volchenko). T.2 Tekhnologiya i oborudovanie (pod redaktsiey V.M.Yampolsiogo). M. Izd. MGTU im. N.E. Baumana., 1996, 574 s.
8. Fromm E-, Gebkhardt E- Gazy i uglerod v metallakh. M. Metallurgiya, 1980, s.499-509.
9. Shorshorov M.Kh., Erokhin A.A., Chernysheva T.A., Kotov GN., Russiyan A.V. Goryachie treschiny pri svarke zharoprochnykh splavov. M.: Mashinostroenie, 1972, s.224.

ABOUT THE AUTHORS

Vladislav Yakovlevich Panchenko *– Institute of Problems of Laser and Informational Technologies of RAS. Director, Professor. Doctor of Physics and Mathematics. An expert in the area of molecular physics, laser physics, the use of lasers for treatment of microelectronic materials and biomedicine.*

Aleksandr Mikhailovich Zabelin *– Institute of Problems of Laser and Informational technologies of RAS. Head of a Department. An expert in the area laser technologies, physico-technical problems of high-power* CO_2 *lasers, laser radiation interaction with matter.*

Anatolii Ivanovich Dedyurin– *Institute of Metallurgy and Materials Science, RAS. Candidate of Technical Sciences, Research Officer. An expert in the area of physico-chemistry of materials for nuclear engineering.*

Irina Valerievna Borovitskaya– *Institute of Metallurgy and Materials Science after A.A. Baikov of RAS. Research assistant. An expert in area of physical metallurgy of high-temperature metals.*

Valentina Vladimirovna Sidorova– *Institute of Metallurgy and Materials. An expert in the area of physical metallurgy.*

COPPER-DIAMOND POWDERS FOR USE IN ELECTRIC CONTACTS

V.V.Ivanov

The class of electrocontact composite materials on the basis the of Cu-Cd-C, system being the prospective substitutes of silver, has been developed. The influence of the type of source powder, the pressing pressure and the sintering temperature on the shrinkage of samples was studied. The optimal process parameters, providing a high level of properties of copper-diamond compositions and the absence of graphitizing of diamonds during sintering, have been determined. Contact-details of new materials are produced in the industrial amount.

INTRODUCTION

Silver has traditionally been the main electrocontact metal, but because of its deficiency and high cost, researchers in the whole world are trying to substitute it with copper possessing many necessary properties. For example, the composite material for discontinuous electrocontacts on the base of cadmium copper, obtained by powder metallurgy methods, containing the addition of fine-dispersed diamond up to 3% [1] is proposed. The tests have shown the quite high level of its working properties in the starting electrical apparatus (contactors, magnetic starters).

Despite the prospects for commercial use of a proposed material, it has been insufficiently studied from the technological and materials science point of view and in the respect of physico-chemical and electrical properties. At the same time, the main working properties of powder material are known to be significantly determined by technology. A some results of investigation of technological properties of powder composition containing 1% of diamond powder: Cu-1Cd-1C (the amount of components is in mass.%) are reported in the present work.

THE SOURCE MATERIALS AND PREPARATION OF SAMPLES

The solution for the problem of forming the dense powder ware with desired properties is the effective mixing of source powder components, the choosing of the scheme and pressing pressure, the regimes and the temperature of the samples sintering, the technique and parameters of additional compacting and the regimes of final heat treatment. The studies were carried out to the powders of electrolytic copper and cadmium, the sprayed alloy Cu-Cd and the explosion diamond. The dispersity of source powders is characterized by size of particles Cu and Cd, being 45 μm (bulk density is 1.7-1.8 g/cm^3); diamond particles size is 1 μm, and sprayed powder alloy particles size is 150 μm at the bulk density of 2.8-3.0 g/cm^3. The use of gross and dense sprayed powders is known to simplify the automatic pressing, sometimes allowing to pass around the mixture granulation and to use press matrix of a smaller sizes. That is why these two main kinds of powders were studied.

Since the structure and morphology of both copper powders are essentially differed, one should expect the difference in their technological behavior. Besides, it has been determined that ingots of electrolytic powder are of higher density after the pressing and the sintering to the equal regimes [2]. In Fig.1, the microscopic pictures of source powders are presented. The particles of electrolytic copper powder have branched dendrite structure and, as sequence, the large surface area. At the same time, the sprayed powder contains a larger particles of irregular shape. The influence of the shape of copper powders particles of different origin on the specific surface S and bulk density d_n was studied in the work [3]. According to the results [3], S for electrolytic and sprayed powders of equal average size is differed, approximately, for 3-5 times. The simplified evaluation due to obvious formula for spherical powder with te average diameter F_{av}~ 50 μm ($S=6/F_{av}d_{Cu}$) yields the value S=140 cm^2/g and can be a some greater, resulting from deviation from regular shape and polydispersity of powder particles. Thus, the specific surface of electrolytic powder is very probably within 600-900 cm^2/g and can be the basis of its heightened activity during the pressing and the sintering processes.

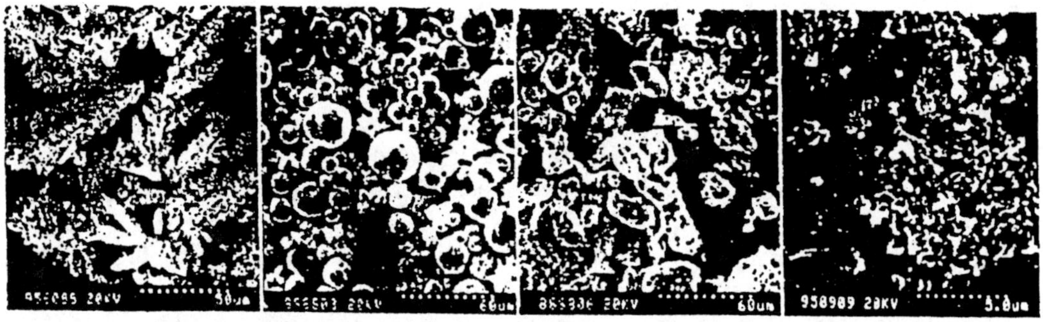

Figure 1. The source powders: a – electrolytic copper, ×600; b – electrolytic cadmium, c – sprayed powder of alloy Cu-Cd, ×500; d – explosion diamond, ×6000.

The powder buttons, corresponding to a given composition, were wetted by isopropyl alcohol and treated in mixer of type "drunken barrel" with the balls as the mixing bodies. The presence of the balls in the mixer was necessary since only in this case the diamond particles are uniformly distributed in the volume, that was reflected on the structure and properties of the finished material. The next experiments to mechanical, electrical and spark-erosion properties have shown a good reproducibility of the measurements on the samples of different groups, that is the evidence of homogenity of powder material and, consequently, the efficiency of mixing process.

The samples of different sizes were pressed from mixtures according to their destination. The pressing pressure was varied within 50-600 MPa, sometimes up to P=1000 MPa. The values P=300-500 MPa, which are usually used during the press forming of copper powders and which provide the necessary porosity of ingots θ 15-20% are within the boundaries [2]. The one-side press static forming in steel mold was used. The little thickness of samples (1-3 mm) along the direction of the pressing, in respect to other sizes, allows to suppose that influence of friction to the walls of mold is negligible, and the pressing is equidense in the volume. In fact, during the further sintering the form of the sample did not change, the uniform change of sizes was observed: the equidense pressing had yielded the equidense sintered sample.

During the whole stages of technology the samples density was controlled by means of measuring their sizes and mass. The hydrostatic weighing technique was used during check measurement to the low-porosity samples. These and other results have coincided within the range of experimental accuracy. The relative density $\varepsilon = d/d_T$ and porosity $\theta = 1-\varepsilon$ was calculated with respect to theoretical porosity d_T of porousless material assuming the additive contribution of cadmium bronze $d=8.89$ g/cm^3 [4] and diamond $d=3.52$ g/cm^3 [5].

Results and Discussion

The experimental results show that the compacting of studied powder mixtures obeys the general pressing behaviors [6,7]. The mixture compacting curves in the coordinates d,P are given in Fig.2. The curve 1 reflects the compacting of mixture on the base of electrolytic powder, and the sprayed one, curve 2 (let us mark for short, respectively: mixture-1, mixture-2).

Although the behavior of curves 1,2 is equal (with increasing the pressing pressure the rate of compacting $d\varepsilon/dP$ decreases), the significant difference in the ability to the compacting of both mixtures is observed. In whole investigated range P, for achieving the equal density the mixture-2 requires approximately for 100 MPa pressure as large. This fact turns to be a natural sequence of both the powder morphology and the internal particles structure. A high-rate consolidation and the hardening of microdrops occur. The cooling rate of microdrops depends on process conditions, reaching 10^2-10^5 K/s [8]. The quickly hardened powder-2 is of microcrystalline structure (there is very low probability of that alloy will become amorphous at the given rates of the cooling) and of the heightened hardness.

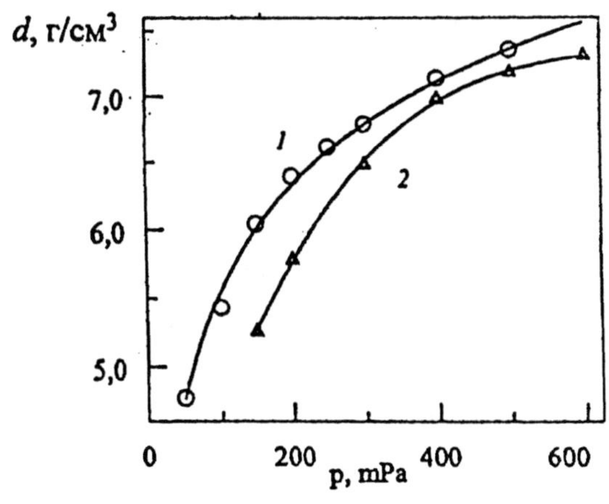

Figure 2. The curves of the compacting powder mixture on the base of electrolytic (1) and sprayed (2) powders.

The powder compacting process is assumed to be divided into three stages according to the differences in the compacting mechanism: 1 – the increase of packing coefficient due to the sliding of the particles regarding to each other, the elimination of arcs and voids; 2 – deformation of near-contact areas; 3 – volume deformation. The transition from first to the second stage occurs at pressures close yield point $\sigma_{0.2}$ or, rather (for metallic powders) $P_{1\ 2}=(1.3\pm0.2)\sigma_{0.2}$ that corresponds to $\varepsilon=0.7$-0.75. The transition to the third stage begins at yielding the materials of particles into pores, i.e. at P HB$_{max}$(HB$_{max}$ is hardness by Brunel of deformed metal) and ε=0.9-0.98. The nonporous material can be obtain at P=(4-8)HB$_{max}$, that is close to the theoretical strength of substance [7]. For the copper HB$_{max}$=878 MPa at 20^{0}C [9] and, thus, the third stage of the pressing could be observed starting from values P 3500-7000 MPa.

The experimental data for every stage of the pressing is usually described by equation

$$\varepsilon = \tilde{\varepsilon}\left(\frac{\tilde{P}}{P}\right)^n \quad (1)$$

where n is constant, ε and P are the values corresponding each to other within the frames of a given compacting mechanism. In logarithmical coordinates this dependence is of linear behavior and allows to calculate the value n:

$$\lg\varepsilon = m + n\lg P \qquad (2)$$

The plotting of our results in the coordinates lgε, LgP has shown their correspondence to the presented equation (Fig.3, curves 1,2). The diagrams are the piecewise-linear dependencies with the break near P=250 MPa for mixture-1 and 300-350 MPa for mixture-2. According to the described conceptions, these pressures correlate with transition from first stage of the compacting to the second one, thus, the value ε=0.73 of the transition is also within typical limits.

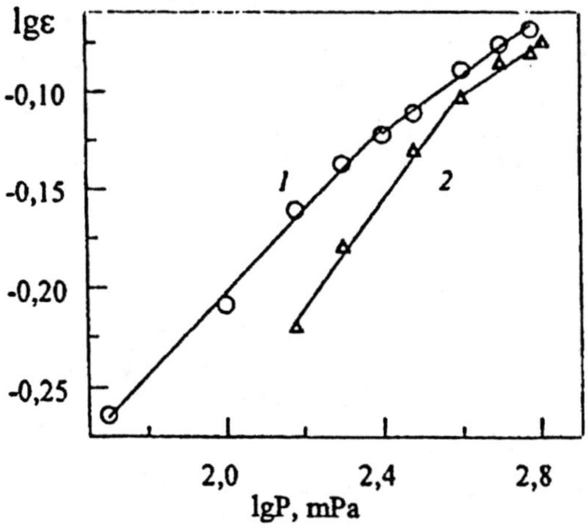

Figure 3. The linear representation of compacting curves.

The comparison of the ability to the compacting of mixture-1 with literature data of the electrilytic copper powder [6,7] shows their complete identity not only to quality, but also to quantity. Line 1 (Fig.3) is described by two equations

ε_1'=0.24P$^{0.21}$ is the part of low pressure, (3)

ε_1''=0.34P$^{0.15}$ is the part of high P. (4)

In the work [7], the exponents n'=0.20, n''=0.15 are presented, that is in good coincidence with our values. The extrapolation of ε_1'' up to P=1000 MPa gives ε=0.93 that also coincides with data of [6]. Such a closeness of the results is not the surprise and is the evidence of weak influence of small additions of cadmium and diamond powders.

The significant differences of linear parts slope in Fig.3 are reflected in the equations analogous to (3) and (4);

$$\varepsilon_2'=0.15P^{0.27}, \quad (5)$$

$$\varepsilon_2''=0.36P^{0.13}. \quad (6)$$

According to exponents $n_2'=0.22$ and $n_2''=0.12$, on the first part of the compacting, the influence of pressure is stronger, and in the second is weaker in comparison to mixture-1. This seems to be reasonable from points of view of compacting mechanisms: the hard particles and particles of regular shape easily response on the change of pressure, sliding relatively to each other, that stimulates to a more dense packing at the first stage. Whereas on the second stage, where the deformation starts playing determinative role, the high hardness of particles causes the inverse effect.

The equation (1) can be presented in other form:

$$P=P_0\varepsilon^{1/n}. \quad (7)$$

The essence of constant P_0 is obvious when assuming $\varepsilon=1$. Then P_0 is the pressure, for which the porousless state is achieved. However, that is right only for the third stage of the pressing under higher pressures and values $\varepsilon>0.9$-0.98. In our experiments this stage is not achieved.

The sintering of the pressed samples was carried out in argon at temperatures 800-960^{0}C during 0.5-1.0 hours. A high temperatures and long duration has essentially complicated the maintenance of a set of chemical composition of the wares. In Fig.4, the diagrams, showing the pressing pressure influence on the relative change of the diameter $\Delta\phi/\phi$ (1), $\Delta h/h$ (2) (marked as $\Delta l/l$) and the density d (3) after the sintering of the samples of the mixtures with electrolytic copper are plotted. It is seen that with increasing the density of the preform the value $\Delta\phi/\phi$ drops arcwise to P 300 MPa. Further, the line is broken and the arcwise dropping of $\Delta\phi/\phi$ with smaller angle slope goes on. Beginning from the values P>400 MPa ($\varepsilon>0.82$), the shrinkage of samples to diameter stops and the expansion starts.

The dependence of the change of sample height is more complicated in the direction of the pressing axis (curve 2). The shrinkage occurs only at the smallest efforts 50-100 MPa. The mutual influence of both linear dimensions leads also to nonmonotomic change of the density: from 6.6 g/cm^3 at 50 MPa to 7.1 g/cm^3 at P=200 MPa with further dropping to 6.7 g/cm^3 at 400 MPa and arising after 500 MPa. Hence, for the variations of P within 100-250 MPa the density changes little: d=7.0-7.1 g/cm^3 [2]. Particularly, it can be caused by insufficient sintering time and by temperature, but the main reasons, probably, are another ones.

The explanation of the behavior of these dependencies should be searched in both the mechanism of the pressing with changing pressure and the complex physico-chemical processes, occurring during the sintering of the pressing.

The break on the curve 1 and the bend corresponding to the break on the curve 2 are close to value P, at which the breaks on Fig.3 are observed and the transition to the second stage of the pressing with the essential particles deformation occurs. Here the interparticle contacts increase and are strengthened, the relative density approaches to ε=0.8, the closed porosity, which is not connected with surface, appears.

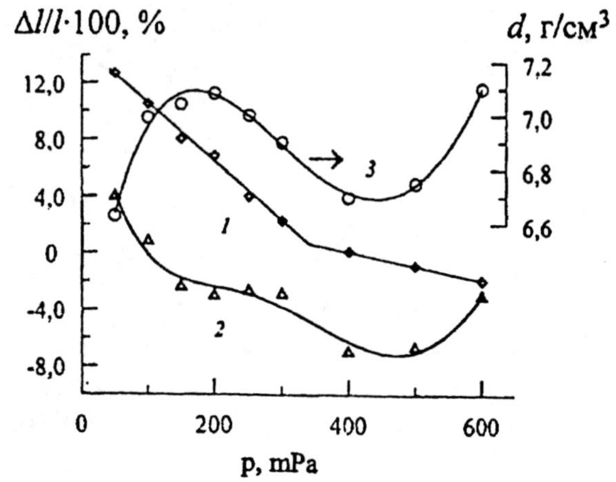

Figure 4. The relative shrinkage (1,2) and change of density (3) of the pressings of mixture on the base of electrolytic copper during sintering (930^{0}C, τ=0.75 h), 1 – diameter, 2 – height.

As it is well known [10], there are several reasons resisting to obtain the high dense wares of copper and its alloys under single pressing and sintering. They have, certainly, the general character: 1 – the tension relaxation, especially in the direction of pressing axis; 2 – the bursting action of gas molded in the pores or created as the result of chemical reactions; 3 – diffusive porosity due to the difference in diffusion coefficients of mixture components or the solution of the addition in the alloy (the alloying).

The action of the second reason in our case is increased by specificity of system. The cadmium, having high partial pressure of the vapor, and the carbon, interacting with copper oxides with creating CO and CO_2, additionally contribute into the disjoining action of gases during sintering and this contribution at the high density of the pressing, when the portion of the closed porosity is great, can be determinative.

The creation of diffusive porosity is especially typical for the systems with hard solutions, that is observed in our system Cu-Cd as well. Diffusion growth during the mutual diffusion of components appears as the result of inequality of partial diffusion coefficients of Cu and Cd in the solid solution, that leads to the appearance of excess vacancies and as the result of their condensation – to diffusive porosity. The preferential diffusion flow goes from components of smaller evaporation heat Q_{evp}, as follows from thermodynamic criterion[11]. In system Cu-Cd the heat of sublimation of components are essentially differed at 298 K Q_{Cu}=337.6 kJ/mol, Q_{Cd} =111.3 kJ/mol [9], so it must be the directed flow of cadmium into copper and the creation of pores on the place of cadmium particles.

The sintering of powder body Cu-Cd can be considered as "with participation of liquid phase". Hence, the components interact between themselves with creating the chemical compounds and solid solutions of cadmium in copper. Taking into account the small amount of the addition (less of the limiting solution of Cd in Cu, being about 2 at.% at temperature of the sintering [16]) the process corresponds to the sintering "with vanishing liquid phase". By A.P.Savitskii [12], in this case the diffusion of liquid cadmium into hard copper occurs, that leads to the increase of dimensions of copper particles, appearance of "arc effect" and the growth of the sintering body.

For excluding the diffusive porosity, it is necessary to determine the reason of its creation, being the use of pure source powders, and to employ the alloying components. This, however, will not completely remove the influence of a given factor in our system: due to the evaporation of cadmium from surface of the ware, the gradients of concentrations, the diffusive flows and, as consequence, the possible diffusive porosity will appear in it.

Because of high partial pressure of cadmium vapor and, consequently, because of possibility of intensive mass transfer in gaseous phase as well as of low cadmium melting temperature, the capillary phenomena, apparently, can significantly influence on the sintering process. According to the Kelvin's formula, with curving the surface the pressure of substance vapor changes for Δp, depending on surface energy γ and curvature of surface (r is radius of curvature): $\Delta p=2\gamma/r$.

Thus, the partial pressure of cadmium is greater upon convex surface, and is smaller upon concave surface, than upon plate. Therefore, inside porous body, the cadmium mass transfer through gas phase from some places into other occurs during the sintering of the pressing.

Liquid cadmium wets copper well: the limiting wetting angle at temperatures about 400^0C does not exceed 20^0 [17]. In this case in the narrow gaps between particles Cu, the concave meniscus of large curvature is created, absorbing the Cd vapors, transferred from plate places. For such a scheme of closed internal cadmium mass transfer the opposite vacancies transfer for compensating the differences in the flows Cu and Cd is organized. The process leads to diffusive washing out of interparticle contacts that is directed against main tendency of sintering. This phenomenon, obviously, can take place in the certain range of temperatures, when atom diffusive mobility in solid phase is high enough.

Besides of general reasons, the significant influence on the shrinkage processes during the sintering is caused by size and shape of particles, the unsoundness of their structure, thus, the origin of powder can change not only quantity results, but also character of dependencies [10].

The diagrams of the relative density change of the pressings of mixture-1 and mixture-2 during the sintering vs. pressing pressure are presented in Fig.5. The beginning parts of these curves are essentially differed. It is seen from diagrams that peak value $\Delta\varepsilon$ for samples of mixture-2 is only 8-9% at P=200-300 MPa, while electrolytic powder is intensively sintered after having been pressed at P=100 and even P=50 MPa ($\Delta\varepsilon$ is 21 and 27%, respectively, is not shown in figure). It is connected with its heightened surface energy and greater number of interparticle contacts.

The behavior of curves is analogous further, the difference is only in absolute values: with increasing P $\Delta\varepsilon$ decreases, dropping to negative areas of unit expansion, but, beginning from certain values P increases again. The increase of $\Delta\varepsilon$ is observed only after ~500 MPa for mixture-1 and after 700 MPa for mixture-2. This increase is connected with the significant strengthening of interparticle contacts and the increase of connectedness of metallic phase, when the porosity decreases to ~20-22% for the pressings (1) and ~15% for pressings (2). As it is seen, in the studied range of sintering temperature 800-960^0C the curves behavior is equal. The experimental points for temperature 800^0C are, basically, between curves 3 and 4 (not shown in figure).

The study has shown that temperature slightly influences on the shrinkage degree. Thus, at P=300 MPa in the range 800-960^0C the shrinkage is 8±1%. However, the stable tendency of decrease of $\Delta\varepsilon$ with transition from 860 to 930 and further to 960^0C is seen.

The sintering of the copper powder pressings or slightly alloyed copper alloys is carried out at temperature 850-1050^0C, but often about 1000^0C, since this makes increasing the porosity, the plasticity and the impact elasticity of material. These increases are connected with low tendency to the grain growth of porous units [2,13]. Thus, for sintering, the maximal possible temperature for a given system is desirable. In a great extent it is determined by the employed equipment and equipment for the sintering, which allow to provide a given chemical composition contact-units.

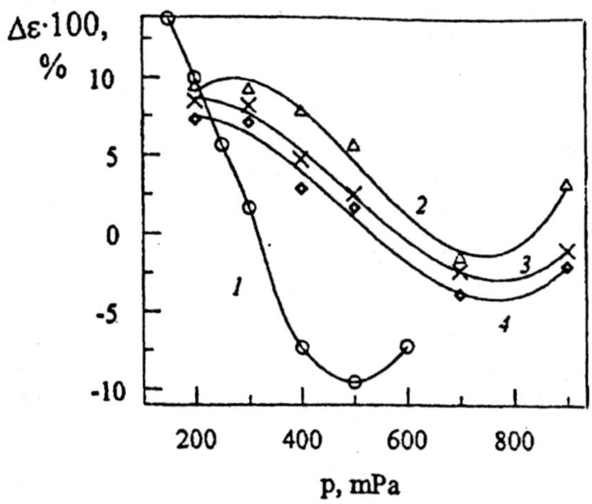

Figure 5. The change of relative density of samples during sintering (τ=0.75 h), 1 – electrolytic powder; 2-4 – sprayed powder. 1,3 – 930°C, 2 – 860°C, 3 – 960°C.

The physico-chemical features of powder alloys of copper are known, as rule, not to allow obtaining quality items with residual porosity less 15% for single pressing and sintering. The additional compacting is necessary, after which the thermal annealing for removing post-pressing tensions follows. These technological stages have not peculiarities in the considered case. For obtaining the material with density $\varepsilon \geq 0.97$ the pressing pressure $P \geq 1100$-1200 MPa is required, and the optimal temperature of finishing annealing is 500-550°C.

The question of possible graphitization of the used fine-dispersed diamond during sintering requires the individual mention. This important question is connected, particularly, with substantiated choice of sintering the temperature and the expediency of using such a diamonds in copper-diamond sintered composites in general. The diamond is known to be metastable under usual conditions. Under low pressures the carbon modification keeps self structure only because of kinetic difficulties of lattice rebuilding process. With increasing the temperature the kinetic limitations are withdrawn, higher of 600°C the diamonds begins interacting with oxygen and after 700-1100°C, according to different authors, the surface and volume graphitization starts. The presence of some impurities makes essentially easy the phase transition as well.

The most complete study of synthetic diamond powder graphitization was, surely, carried out in work [14]. The powders of different fractions from 1/0 to 125/100 were studied (the numerals correspond to maximum and minimum limits of particles size, μm) in vacuum (pressure of residual gases is 0.1 Pa), within temperature range 1600-2100 K. The extrapolation of results to the temperature close to our T_{SP} gives very insignificant rate of surface graphitization about $6 \cdot 10^{-10}$ g/cm^2·h. However, under real conditions of sintering the copper-diamond compositions when a diamond contacts with a metal, with a copper oxygen and with an atmosphere containing oxygen, the situation can be different. The interaction of diamond with an oxygen or an oxygen-containing compounds leads to increasing the number of defects on the surface, and the presence of defects makes easy the process of graphitization, since the latter goes with increasing the volume [15].

Therefore the special experiments on studying the possibility of graphitization of "explosive" and static synthetic diamonds under conditions of laboratory and industrial process of the sintering of copper-diamond pressings have been carried out. The samples after sintering were dissolved, the solid dredge was extracted, which was analyzed by means of X-ray analysis technique further. The results show that in the indicated ranges of conditions the graphitization was not observed.

CONCLUSION

The new class of electrocontact composition materials on the base of system copper-cadmium-diamond, being the prospective substitute of silver, has been developed. The influence of type of source powder, the pressing pressure and the sintering temperature on th shrinkage processes have been studied. The pressure influences in complex way with increasing the density of the pressing the shrinkage of samples is substituted with the expansion. The influence of temperature within studied range 800-960^{0}C is relatively large. The optimal technological parameters, providing high level of properties of copper-diamond compositions and absence of graphitization of diamonds during sintering, were defined. These parameters are essentially depending on source powder materials.

Contact-units of the studied material in the composition of the starting electrical equipment have been tested in China on the correspondence to national standard CJ-10 and are presently manufactured in the industrial scale on the joint-stock company.

The author is grateful to professor O.N.Breusov for the diamond powders presented for the study.

REFERENCES

1. Patent Rossii no.2073736. Spechennyy elektrokontaktnyy material na osnove medi. Ivanov V.V., Kirko V.I., Ivanov Vl.Vl.
2. DovydenkovaA.V.,RadomyselskiyI.D. Poroshkovaya metallurgiya, 1982, no.3, s-44-53-
3. Balasubramanian V., Singh P., Ramakrishnan P. Powder Met. Intern., 1984, v.l6, no-2, p.56-59.
4. Smiryagin A.P., Smiryagina N.A., Belova A.V. Promyshlennye tsvetnye metally i splavy. M.:Metallurgiya, 1974, 488 s.
5. Bokiy G.B., Bezrukov G.N., Klyuev Yu.A., Naletov A.M., Nepsha V.I. Prirodnye i sinteticheskie almazy. M.: Nauka, 1986,222s.
6. Balshin M.Yu. Nauchnye osnovy poroshkovoy metallurgii i metallurgii volokna-M.: Metallurgiya, 1972, 335s.
7. Andreeva N.V., Radomyselskiy I.D., Scherban N.I. Poroshkovaya metallurgiya, 1975, no.6, s.32-42.
8. Amorphous metallic alloys. Ed. I.Luborsky. London: Butterworths, 1983, 534r.
9. Svoystva elementov. Spravochnik, t.1, pod red. G.V.Samsonova. M.: Metallurgiya, 1976, 599 s.
10. Andrievskiy R.A. Poroshkovoe materialovedenie. M.: Metallurgiya, 1991, 203 s.
11. GeguzinYa.N. Fizika spekaniya. M.: Nauka, 1984,312s.

12. Savitskiy A.P. Zhidkofaznoe spekanie sistem s vzaimodeystvuyuschimi komponentami. Novosi-birsk: Nauka, 1991, 183s.
13. Eudier M. Powder Met., 1978, no.2, p. 101-104.
14. Fedoseev D.V., Bukhovets V.L., Vnukov S.P., Anikin B.A. Poverkhnost, 1986, no.1, s.92-99.
15. Uspenskaya K.S., Tolmachev Yu.N., Fedoseev D.V. Zhurn. fiz. khimii, 1982, t. 56, no.2, s.495-496.
16. Vol A.E., Kogan I.K. Stroenie i svoystva dvoynykh metallicheskikh sistem, t.4. M.: Nauka, 1979, 576 s.
17. Khludov E.A., Pashkina G.V., Belyy D.I., Izrailev A.E., Zhuykov V.P. Kadmirovanie mednoy provoloki i puti povysheniya kachestva metallopokrytiy. Adgeziya rasplavov i payka materialov, 1990, vyp-23, s.94-98.

About the Author

Victor Vladimirovich Ivanov– *Research Physico-Technical Institute of Krasnoyarsk State University. Assistant professor, Candidate of Chemical sciences. An expert in area of powder materials science.*

LASER-ULTRASONIC HARDENING OF STEEL SURFACE

D.M. Gureev

The possibilities of laser-ultrasonic hardening of steel surface under the controlled change of structural-stressed state in the surface layer have been studied. The advances of laser-ultrasonic treatment with respect to laser treatment in creating a harder and a deeper zones of the hardening surface with temporal controlling the structure-phase composition formation process and the level of residual stresses are demonstrated.

INTRODUCTION

The use of laser radiation as the surface heat source is in the basis of changing the structural-stressed state of the surface under laser exposure. Thus the problem of controlling the formed structural-stressed state is one of the main problems for the practical realization of laser heat-treatment of surface of machine units and instrument with the aim to increase their wear resistance. One can propose several way of solving this problem, but they all somehow or other are reduced to using the integrated approach, based on the general physical conception of energetic similarity in the behavior of open system "metal-external effect" under heat and mechanical injection of energy and of the beginning of non-equilibrium phase transformation at reaching the critical value of energy density in the metal irrespective of the reaching method. One of such complex approaches is the realization of the consequence: the plastic deformation – laser heat treatment of the surface. Among the methods of plastic deformation there are: the rolling by the balls or rollers, the shot-blast treatment, the ultrasonic instrument treatment. The latter method is the most effective due to its high-cyclicity, as well as it is easily built into technological processes of high-concentrated energy flow treatment of the surface. Therefor the use of the combined laser-ultrasonic source seems to be prospective, realizing the influence on metallic surface of ultrasonic vibrations (mechanical energy) and laser radiation (heat energy).

During the laser-ultrasonic hardening the input of ultrasonic mechanical energy must change the kinetics of phase transformation and diffusion processes and effect on the degree of their completeness under conditions of high rates of the laser heating and the cooling. This is in the base of flexible operation of the structural-stressed state of the treated surface.

The aim of present work is to determine experimentally the possibilities of laser-ultrasonic hardening of the high-chromium ledeburite steel X12M under controlling change of its structural-stressed state.

THEORETICAL PREREQUISITES FOR LASER-ULTRASONIC HARDENING OF SURFACE

The temperature of the beginning of α-γ-transformations in the steels (A_1) is known to decrease after their plastic deformation [1,2]. The effect is explained by possible relaxation of part of energy of cold hardening during phase transformation, according to which the equality of thermodynamic potentials of phases, determining the phase equilibrium temperature, is achieved at lower temperatures. In our work [3], on the base of conception of the matter limiting state in the point of phase change and the use of dilatation of the lattice as the measure of the energy stored by the material, the expression for cold-hardening energy ΔU/ΔT, the relaxation of which leads to decreasing the temperature A_1 for one degree, has been obtained

$$\frac{\Delta U}{\Delta T}=\frac{(\alpha_\alpha+\alpha_\gamma)GV}{2} \quad (1)$$

Here α_α and α_γ are the coefficients of linear thermal expansion of α and γ phases, respectively, G is modulus of rigidity, *V* is specific volume. The use of a known values for α_α, α_γ, *G* and *V* [2,4] gives the value $\Delta U/\Delta T$ 6.7 J/mole·deg [1] that is in good agreement (with accuracy of 20%) with experimental values of 5.6 J/mole·deg [1].

The phase equilibrium temperature shift due to plastic deformation is defined by expression [3]

$$\Delta T=\frac{\rho\alpha^2}{\alpha_\alpha+\alpha_\gamma} \quad (2)$$

where ρ is density of dislocations, α is lattice parameter. According to a given expression, for typical values of the limiting density of dislocations $\rho=(2\text{-}4)\cdot10^{12}$ cm^{-2} under plastic deformation the maximal possible temperature shift A_1 does not exceed 50-100 degree. Thus, the primary plastic deformation of surface allows to expand the region of α-γ-transformations in steel up to 100 degrees.

A some techniques of plastic deformation are known as well [5]. First of all, one should include the technique of plastic deformation of surface with balls or rollers. In this case the deforming element as the ball or roller with force of F is pressed to the surface of the treated detail which moves with the given velocity *v*. For all the advances, because of relatively small pressures and deformation rates, a given technique does not allow to use the ability of metal to the hardening enough.

A more complete use of reserve of the increase of metal hardness is realized by replacing the static method of plastic deformation with the impulse method. The embossing method, the shot-blast hardening treatment method, as well as the most effective method of treatment of surface by ultrasonic device are among these methods.

The main parameters of ultrasonic treatment, determining the quality of surface layer and, consequently, the service properties, are the vibration amplitude ξ, the die hold-down pressure *F* and vibration frequency *f*. These parameters determine the value of the working loads and the multiplicity of single acts of surface plastic deformation. Together with them the quality of surface layer is also determined by the movement velocity *v*, by the rounded radius of instrument R and by roughness of surface R_z.

It is reliably determined [5], that extent of plastic deformation under impact exposure of ultrasonic instrument is determined by value of peak load F_p, which is not arbitrary ultrasonic parameter but depends on amplitude of vibration ξ, die hold-down pressure F and characteristics of the treated material

$$F_p \ (9\pi^2 k\xi F^2)^{1/2}, \qquad (3)$$

where k is the coefficient characterizing the mutual resistance of the treated material and material of ultrasonic instrument during their interaction. The peak loads are much higher of their die hold-down pressure. The increase of efficiency of plastic deformation of surface by ultrasonic instrument is the sequence of that. The multiplicity of deformation, which is typical for ultrasonic treatment, significantly decreases the initial roughness of the surface and leads to the increase of maximal depth of deformed layer h

$$h \sim \left(\frac{F\xi}{H}\right)^{1/2} \qquad (4)$$

where coefficient H characterizes the plasticity of the deformed material. A given coefficient depends on the load cycle number. Under multiple load, being typical for ultrasonic exposure, every repeated load should be considered as the deformation of material with the high hardness. Thus the change of hardness in the limited range of the load multiplicity N is described by functional dependence

$$H=H_0\ln(\beta N), \qquad (5)$$

where H_0 is the hardness under single exposure, β=6.3.

Thus, due to the change of roughness and structure of surface the ultrasonic treatment in zone of laser exposure must effect on change of surface absorption ability and thermal and physical characteristics of material. This, together with the change of kinetics of phase transformations and diffusion, must develop in changing the geometric parameters and micro-durometer characteristics of zone of laser effect. In particular, in work [6] the increase of microhardness of surface during its treatment by laser together with laser ultrasound was observed.

MATERIAL AND EXPERIMENTAL TECHNIQUE

The semi-heat-resistant ledeburite high-chromium instrumental steel X12M in the initial annealed state was used as the investigated material. The chemical composition of steel (in mass%) is 1.45-1.65C; 11.0-12.0Cr; 0.40-0.6Mo; 0.15-0.30V; 0.15-0.40Mn; 0.15-0.35Si; the rest is Fe. The main carbide phase in a given steel is the complex chromium carbide $(Cr,Fe,Mo,V)_7C_3$.

The surface of samples of dimension 30×170 mm was undergone to laser-ultrasonic hardening. The thickness of samples was 4 mm.

The ultrasonic plastic deformation of the surface was performed by using the ultrasonic generator UZG1-1 and magnetostrictive converter PMS1-1. The frequency of ultrasonic vibration f was 22 kHz. The concentrator of half-wave length with the rounded radius of R=5 mm was attached to the end of magnetostrictive converter. The concentrator of steel 40X in the initial annealed state and after quenching at 840^0C in the water was used. The die hold-down pressure of concentrator to the surface of sample was F=50 N. The amplitude of ultrasonic vibrations ξ has been changing from 5 to 25 μm. The concentrator has been moving over the surface of sample along the direction of initial polishing with the velocity of v=0.5 and 4.0 mm/s. In the realized conditions of the experiment the parameter $p=kF\xi/v$, characterizing the efficiency of plastic deformation of the surface by ultrasonic instrument technique, was varied within (5-30)·10^{-2} N·s. In the expression for parameter p the dimensionless coefficient k takes into account the relation of initial hardness of deformed material and concentrator material and engagement factor of the individual passes during ultrasonic treatment.

For the surface laser heat-treatment the pulsed solid-state laser YAG:Nd Kvant-15 (the radiation wavelength λ=1.06 μm) was employed. The laser radiation was focused into the spot of diameter d=1 mm. The pulse energy E changed from 3.5 to 6.5 J that for pulse duration of t 5 ms provides the realization of treatment regime both without fusion and with fusion of the surface. Laser treatment was performed by single pulses with engagement factor of individual spots of 0.2.

The geometric parameters, micromechanical characteristics and structure-phase composition of zones of ultrasonic and laser-ultrasonic treatment were investigated with using the methods of metallography, micro-durometry and X-ray phase analysis. For etching the transversal ream of samples the solution of nitric acid of 4% in ethyl alcohol was used. The microhardness was measured on the device PMT-3 under load of 0.98 N. The recording of difractograms was performed to diffractometer DRON-2 with using K_α-radiation of iron. Thus the amount of residual austenite was determined by method of homology pair.

The residual tensions in surface layers of samples were measured by X-ray method "$\mathrm{Sin}^2\varphi$" to the portable plant Tensionflex PSF-2M for four angles φ=0, 10, 20 and 30 deg. In range of angles 2θ=140-170 deg. in a given case the chromium was the source of X-ray radiation.

EXPERIMENTAL RESULTS AND DISCUSSION

An obtained results are in full agreement with theoretical prerequisites of influence of laser-ultrasonic treatment on geometric parameters and micro-durometry characteristics of zones of laser exposure. The corresponding data are presented in Fig.1. From comparison of diagrams of distribution of microhardness $H_{0.98}$ in depth it follows that laser-ultrasonic treatment method allows to form a harder and deeper surface layers at smaller values of laser pulse energy E. In general, the surface ultrasonic plastic deformation leads to decreasing of the limiting laser pulse energy at which the melting temperature of metal is achieved. Thus for parameter p 0.3 N·s this reduction of limiting energy of laser pulse is 20-25%. The change of thermal and physical characteristics of surface layer resulting from plastic deformation by ultrasound is in the base of change of parameters of laser source under laser-ultrasonic exposure.

For evaluating the α-γ-transformation temperature shift under laser-ultrasonic exposure, which appears in increasing the hardening layer depth, let us use the expression deduced from simple formula obtained in work [7], which binds the hardening depth with the parameters of pulse laser source, optical and thermal and physical characteristics of material. In expression (6), a_T is thermal diffusivity, T_0 is surface temperature, T is temperature to which the hardening layer depth z corresponds, e is the base of natural logarithm. The evaluation shows that as provided by theoretical prerequisites the ultrasonic plastic deformation of surface leads to the shifting of temperature of the beginning of α-γ-transformation into zone of laser exposure for value, which does not exceed 100 deg.

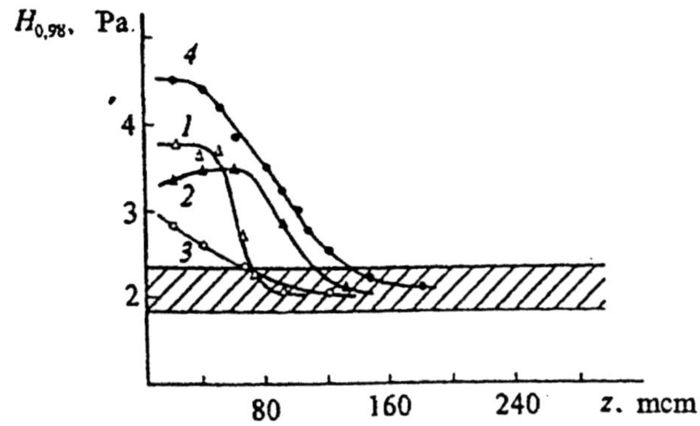

Figure 1 Distribution of microhardness $H_{0.98}$ vs. depth z under laser (1,2), ultrasonic (3) and laser-ultrasonic (4) exposure. 1 – treatment without surface fusion, the energy of laser pulse E=6.0 J; 2 – treatment with surface fusion, E=6.2 J; 3 – 0; 4 – treatment without surface fusion, E=4.8 J. Parameter p is 0 for 1 and 2; $29.4\cdot10^{-2}$ N·s for 3 and 4.

Ultrasonic plastic deformation of surface assists in passing more completely the process of austenization under high rate of laser heating. As the result, the greater amount of residual austenite remains in zone of laser-ultrasonic exposure under other equal conditions than in zone of laser exposure. The characteristic fragments of difractograms of surface layers of zones of laser, ultrasonic and laser-ultrasonic exposure are presented in Fig.2. These fragments of difractograms clearly demonstrate the extent of influence of ultrasonic plastic deformation on the change of quantitative ratio between main phases of solid solution, being formed under laser exposure.

In addition to the mentioned above, the generalized results of X-ray analysis as the dependence of amount of residual austenite A_0 in surface layers of laser and laser-ultrasonic exposure zones, respectively, vs. laser pulse energy E (Fig.3) and vs. parameter p are given in Fig.3 and 4. Here the diagrams of change of longitudinal σ_y and transversal σ_x residual tensions, respectively, to the direction of polishing, being formed in the surface layers under laser, ultrasonic and laser-ultrasonic exposure depending on laser pulse energy E (Fig.3) and parameter p (Fig.4). The analysis of the residual tensions, which are formed, shows that ultrasonic treatment, as well as the laser, decreases φ-splitting of residual tensions, determined by conditions of initial polishing of surface and caused by inhomogeneous depth distribution of dislocation [8]. In surface layer during its plastic deformation by ultrasound the residual compressions, increasing with parameter p, are formed. The laser treatment leads to formation of residual strains that is caused by primary distribution of thermal component, determined by large (10^4-10^5 deg/mm) temperature gradient under pulse exposure. The contribution of phase component, which is responsible for formation of residual compressions, depends on the level of development of martensitic transformations and degree of tetragonality of formed martensite. With increasing the amount of residual austenite the martensite amount decreases, but high-carbon martensite fraction increases in its general content, therefore as far the austenization process develops the contribution of phase component can exceed the contribution of thermal component. This is in the base of controlling the structural-stressed state of surface by laser-ultrasonic source and is confirmed by a curve of dependence of residual tensions σ vs. parameter p with maximum at $p=22\cdot10^{-2}$ N·s, given in Fig.4.

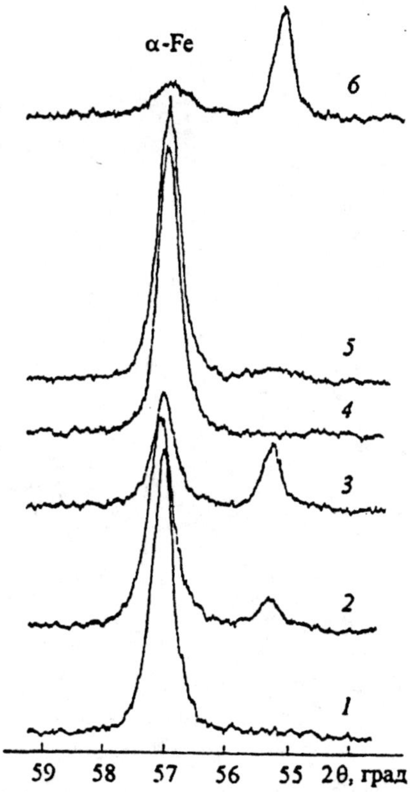

Figure 2 Fragments of difractograms of initial surface (1) and zones of laser (2,3), ultrasonic (4) and laser-ultrasonic (5,6) exposure. The laser pulse energy E, J: 1,4 – 0; 2 – treatment without surface fusion, E=6.0 J; 3 – treatment with surface fusion E=6.2 J; 5,6 – treatment without surface fusion, E=6.2 J. Parameter p is 0 for 1-3; $29.4 \cdot 10^{-2}$ N·s for 4 and 6;$4.9 \cdot 10^{-2}$ N·s for 5.

Fig.5 exhibits the degree of homogeneity of the forming residual tensions σ along direction of treatment of surface by ultrasonic (+y) and laser-ultrasonic (-y) source. The latter is important from technological point of view for practical realization of laser-ultrasonic hardening of surface.

It should be noted that laser-ultrasonic hardening of surface is not limited by using only pulse laser sources. The main results of a given work can be put in the base of developing the technological processes of laser-ultrasonic treatment of surface on the base of high-power technological continuous action lasers. The necessary condition, which is desirable to follow, is in choosing such a parameters of ultrasonic and laser exposures, at which the depths of zones of plastic deformation and laser hardening are commensurable. The time between ultrasonic exposure and subsequent laser exposure can be arbitrary, since the relaxation of deformed state at room temperature is not limited by real time interval.

While using the technological continuos action lasers, the realization of process of thermomechanical treatment of surface, consisting of the plastic deformation of heated zone of laser exposure by ultrasonic instrument, becomes additionally possible. Thus the temperature of areas, undergone to plastic deformation, under other equal conditions will be determined by distance between spot of the focusing laser radiation and area of touch of surface with concentrator of ultrasonic instrument.

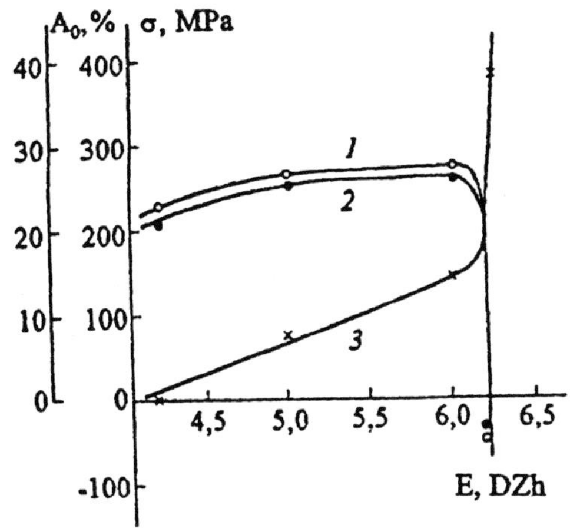

Figure 3 Longitudinal σ_y (1) and transversal σ_x (2) residual tensions and amount of residual austenite A_0 (3) in surface layers of laser exposure zones depending on energy of laser pulse E.

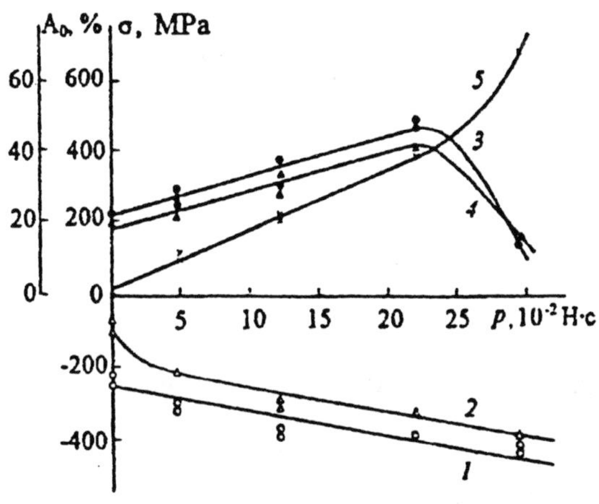

Figure 4 Longitudinal σ_y (1,3) and transversal σ_x (2,4) residual stresses and amount of residual austenite A_0 (5) in surface layers of zones of ultrasonic (1,2) and laser-ultrasonic (3-5) exposure depending on parameter p. Laser pulse energy E, J: 1,2 – 0; 3-5 – 4.2, treatment without surface fusion.

Conclusion

1. It has been experimentally demonstrated that under laser-ultrasonic exposure a harder and a deeper zone of surface hardening at less laser pulse energy is formed.
2. The plastic deformation of surface by ultrasound, influencing on the passing completeness of process of austenization, allows to control effectively the structural-stressed state of surface even under pulse laser exposure.
3. Laser-ultrasonic exposure provides the homogeneous enough distribution of residual stresses over treated surface that has a more practical significance for solving the problems of laser technology.

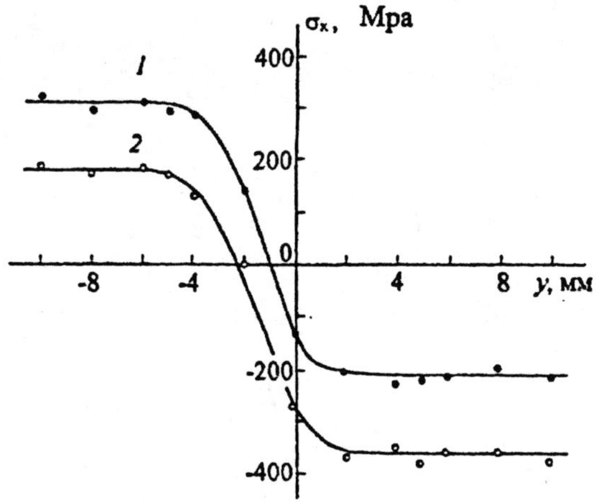

Figure 5 Distribution of transversal residual stresses σ_x along surface, treated by ultrasonic (+y) and laser-ultrasonic (-y) source. Parameter p, N·s: 1 – 4.9·10^{-2}; 2 – 29.4·10^{-2}; Laser pulse energy E=4.2 J, treatment without surface fusion.

REFERENCES

1. Gridnev V.N., Meshkov Yu.Ya., Oshkaderov S.P., Trefilov V.M. Fizicheskie osnovy elektrotermi-cheskogo uprochneniya stali. Kiev, Naukova Dumka, 1973, 334s.
2. Dyachenko S. S. Obrazovanie austenita v zhelezouglero-dietyh splavah. M.: Metallurgiya, 1982, 127 s.
3. Mednikov S.I., GureevDM. PismavZhTF, 1992, t.18, vyp. 5, s.25-27.
4. Kikoin M.K. Tablitsy fizicheskih velichin. Spra-vochnik. M.: Atomizdat, 1976, 1008 s.
5. Abramov O.V., Dobatkin V.I., Kazantsev V.F., Kulemin A.V., Nekrasova S.Z., Panov A.P., Pugachev S.I., Semenova N.G., Statnikov E.Sh., Eskin G.I. Voz-deystvie moschnogo ultrazvuka na mezhfaznuyu poverhnost metallov. M.: Nauka, 1986, 278 s.
6. Sindeev V.I., Ishakova G. A. Fizika i himiya obrabotki materialov, 1988, no.2, s.59-64.
7. Gureev D.M., Katulin V.A., Nikolaev V.D., Petrov A.L., Yaldin Yu.A. Fizika i himiya obrabotki materialov, 1985, no.2, s.22-25.
8. Gureev D.M. Fizika i himiya obrabotki materialov, 1994, no.2, s. 102- 111.

ABOUT THE AUTHOR

Gureev Dmitriy Mikhailovich – *Samara Branch of Physical Institute after P.N.Lebedev of RAS. A professor, doctor of Physics and Mathematics, principal research officer. An expert in area of solid state physics, laser treatment of materials.*

MECHANISMS OF ULTRASONIC PRESSING OF CERAMIC NANOPOWDERS

O.L.Khasanov, E.S.Dvilis, Yu.P.Pokholkov and V.M.Sokolov

The propagation of ultrasound through superdispersed powder (SDP) ZrO_2-Y_2O_3 and possible effects of ultrasonic exposure while pressing SDP, such as destruction of agglomerates, activation of SDP particles under acoustic streams; the dense pressing of SDP with preserving the superdispersed (nano) structure of the pressed billet are analyzed. It has been shown that result of ultrasonic exposure is determined by pressure P and ultrasonic intensity I, the values of which are separated onto the ranges by critical parameters P_c and I_c. For $P=P_c$ the SPD turns from gas-dispersed state into solid porous state, the I_c characterizes the threshold of acoustic stream emergence in powder body.

INTRODUCTION

The superdispersed (nano) powders are used in powder technology for creating nanoceramics and other nanocrystalline materials, in which the unusual physico-chemical and high running abilities are caused by submicron and nanometer sizes of crystallites [1,2].

The thermodynamic metastability of SDP allows to synthesize the unique compounds and phases, activating the most important stages of process due to chemical, surface activity of SDP. However, these factors cause their shortages for powder metallurgy: the agglomeration, the sorbed impurities, the dusting, the bad compressibility and formability due to low bulk density. The agglomerates can be bound with both the strong ion-covalent bonds and the weak Van der Waals forces. For example, the density and the grain of zirconium ceramics essentially depends on the agglomerate size d_a of source SDP ZrO_2-Y_2O_3; high-dense ceramics with ρ_c>98% can be produced provided that d_a<1 μm [1].

The nanostrucure of ceramics is known to be formed when the conditions for homogeneous nucleation with large number of submicron nucleuses are provided. Therefore it is important to retain the nanostructure of the pressed billet prior to sintering, preventing the plastic deformation of SDP during the pressing, i.e. pressing the powder provided that elastic contacts of SDP particles.

For overcoming these problems the special methods of SDP treatment are required, the ultrasonic (US) exposure being one of those [3,4].

In a given work the possible effects under ultrasonic exposure over superdispersed ceramic powder ZrO_2-5.4 mass.% of Y_2O_3 during its uniaxial pressing are analyzed.

ANALYSIS

The dependence of sound speed c in powder body vs. pressing pressure P in the case of elastic contact of particles is as follows [5]:

$$c(P)^2 = \left[\frac{3\pi}{4}P\left(1-\nu_0^2\right)\right]^{1/3} \frac{E_0^{2/3}\rho_{OTH}^{n-5/3}(1-\nu)}{(1+\nu)(1-2\nu)\rho_0} \quad (1)$$

Here

$$n = \frac{\ln(\beta E/E_0)}{\ln \rho_{OTH}} \quad (2)$$

$$E(P) = \frac{\Delta P h}{\Delta h} \quad (3)$$

$$\beta = \frac{2x}{d} = \left[\frac{3\pi P\left(1-\nu_0^2\right)}{4E_0\rho_{OTH}^2}\right]^{1.3} \quad (4)$$

$$\nu = \nu_0 \rho_{OTH} \quad (5)$$

In (1)-(5), ν_0 and E_0 is Poisson's ratio and Jung's module of dense material; $\nu=\nu(P)$ and $E=E(P)$ are current values of these parameters in the pressed powder body; ρ_0 is density of material of SDP particles; $\rho_{rel}=\rho/\rho_0$ is relative density of the pressed powder; ρ is its current density; Δh is the pressing height reduction of h at corresponding increase of the pressure of ΔP, x is the average radius of the area of elastic contact of two particles SDP of diameter of d.

The main parameters of US-exposure over SDP are defined from a known relations [6,7]:

the oscillation shift speed of powder particles at US-frequency of $\omega=2\pi f$

$$\upsilon=\omega\xi=2\pi f \quad (6)$$

the sound pressure amplitude on the powder particles

$$p=\upsilon\rho c; \quad (7)$$

the intensity (power density) of US-vibrations (USV)

$$1 = \frac{\rho c(\omega\xi)^2}{2} = \frac{p\upsilon}{2} \quad (8)$$

the acoustic resistance of powder

$$z=\rho c; \tag{9}$$

the attenuation constant of US-vibrations in powder is

$$\alpha = \frac{\pi \mu f}{c} \tag{10}$$

$$K = \frac{E}{3(1-2\nu)} \tag{11}$$

In (6)-(11) ξ is oscillatory shift of SDP particles, μ is internal friction, K is the module of bulk elasticity (uniform pressing) of powder body.

The acoustic streams in SDP are absent, when the oscillation shift of powder nanoparticles under ultrasonic exposure ξ does not exceed their average size d:

$$|\xi| \leq d \tag{12}$$

In this case, according to (6), at f=20 kHz the oscillation speed of SDP particles of dispersity of d=100 nm does not exceed the value $\upsilon=2\pi f\xi=2\pi fd=1.26\cdot10^{-2}$ m/s.

The critical intensity of USV, to which the acoustic streams are absent and the condition (12) is true, is defined from (8):

$$I_c = \frac{\rho c(\omega d)^2}{2} \tag{13}$$

In (1)-(13) the Jung's module E and sound speed c, the ability to the densification $\rho(P)$ in SDP depend on powder dispersity d [2,3,5] and pressure P. Therefore during the pressing of powder under US-exposure the sound speed c and parameters p, I, z, α, K are in complex dependence on d and P, i.e. the dispersity of the pressed powder essentially effects on absolute values of the above parameters.

Let us determine the appearance of dependence of speed and attenuation of high-power ultrasound in SDP ZrO_2-5.4% Y_2O_3 with d 100 nm during its pressing, issuing from experimentally confirmed densification curves $\rho=\rho(P)$.

Experiment

The uniaxial pressing was performed in steel cylindrical mold. The US vibrations were supplied along radial direction to the mold matrix from magnetostrictive converter [8]. The frequency of USV was determined by resonance frequency of whole acoustic system, being of 21.5 kHz. The level of ultrasonic exposure over the pressed powder was defined by the electrical power output of US-generator of W=0.25-3 kW. The electrical power utilization factor in the employed acoustic equipment was about 30%. In the case of usual static uniaxial pressure W=0.

The pressure pressing P was increased up to value 744 MPa. The internal diameter of matrix was D=10 mm, the SDP button mass M=0.65 g. The plasma chemical SDP ZrO_2-5.4%Y_2O_3 with the average diameter of source particles of d=150 nm was investigated.

The densification curves were calculated due to results of consecutive measurement of the shift of the stem Δh_i to the clockwork micrometer IC-10 while increasing the pressing pressure ΔP_i from 0 up to 744 MPa:

$$\rho_1(P_i) = \frac{4M}{\pi D^2 H_i(P_i)} \quad (14)$$

where $H_i(P_i)=H_{i+1}(P_{i+1})+\Delta h_i-\Delta h_{iE}$ (15)

The curves were built from finite value of the pressed billet height $H_{i+1}(P_{i+1})=H_k(P_{max})$. The influence of elastic deformations of equipment Δh_{iE} was taken into account in the calculation.

The measurement accuracy of Δh is 10 μm, ΔP's one is 6 MPa. There were pressed at least 4 samples under identical conditions for every USV power level W and in the case of static pressing without the USV (W=0). The obtained densification curves are the averaging of the above experimental data.

After having taken out the pressed billets from moulds, their diameter D_p was measured and the elastic aftereffect to the diameter has been determined. The latter parameter characterizes the level of internal tensions in the pressing:

$$\delta = \frac{100(D_p - D)}{D_p} \quad (16)$$

All the compressed samples were sintered simultaneously in the air at 1650°C.

In Fig.1, in semi-logarithmic scale the fragments of densification curves $\rho=\rho(P)$ in the region of small pressures are presented. For every curves, the typical flex point is observed (with corresponding value of some critical pressure P_c), which in [9] for the case of static uniaxial pressing is characterized as the point of transition from the powder particle redistribution regime to the particle deformation regime. In the range P>100 MPa the dependencies are monotonous ones and are practically linear in a given scale (the incut in Fig.1).

The value P_c for every densification curve in the range P=6-50 MPa was defined from condition

$$\frac{d\rho(P)}{dP} = \left[\frac{d\rho(P_c)}{dP}\right]_{max} \quad (17)$$

The change of P_c depending on W is shown in Fig.2.

In Fig.3 the dependencies of bulk elasticity module $K(P)$ of the pressed powder are exhibited, which are calculated due to (11), (3), (5), (14).

Due to (1)-(5), (10) and experimental data of $\rho=\rho(P)$ the dependence family $c(P)$ and $\alpha(P)$ under different levels of US-power W have been calculated. The coefficient of internal friction for SDP ZrO_2 (d=100 nm) is equal to μ=1.5 [3]. The results of calculations for W=1 kW are presented in Fig.4.

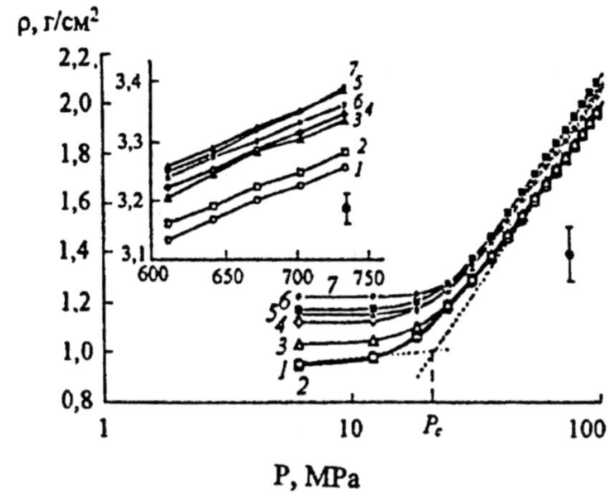

Figure 1 The densification curves of SDP ZrO_2-5.4%Y_2O_3 under static pressing without US-exposure (1) and under US-exposure of different power *W*, kW: 2 – 0.25; 3 – 0.5; 4 – 1; 5 – 1.5; 6 – 2; 7 – 3.

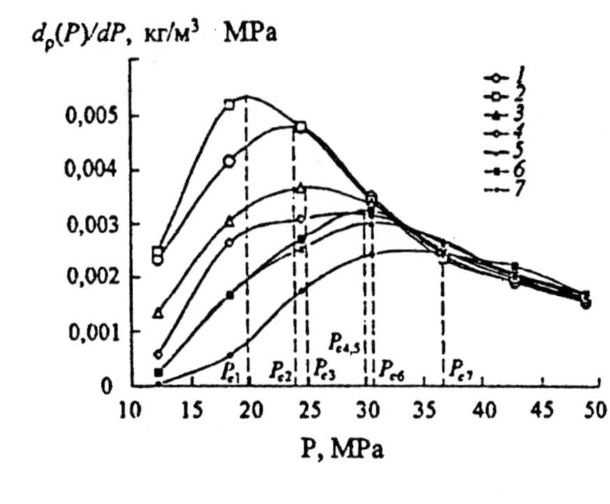

Figure 2 Critical pressure P_c for different regimes of US-pressing, kW: 1 – 0; 2 – 0.25; 3 – 0.5; 4 – 1; 5 – 1.5; 6 – 2; 7 – 3.

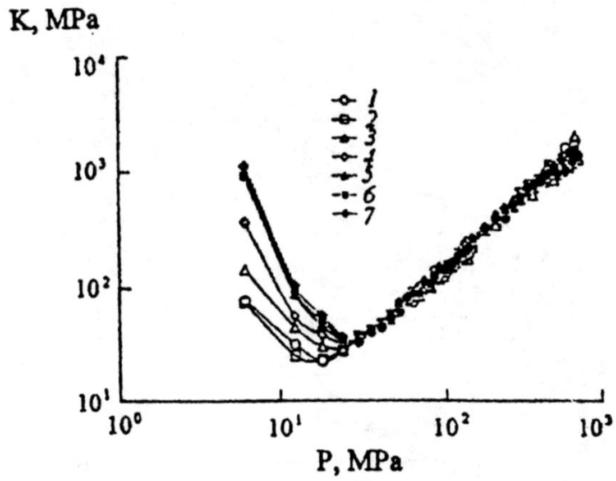

Figure 3 Bulk elasticity $K(P)$ of the pressed powder, kW: 1 – 0; 2 – 0.25; 3 – 0.5; 4 – 1; 5 – 1.5; 6 – 2; 7 – 3.

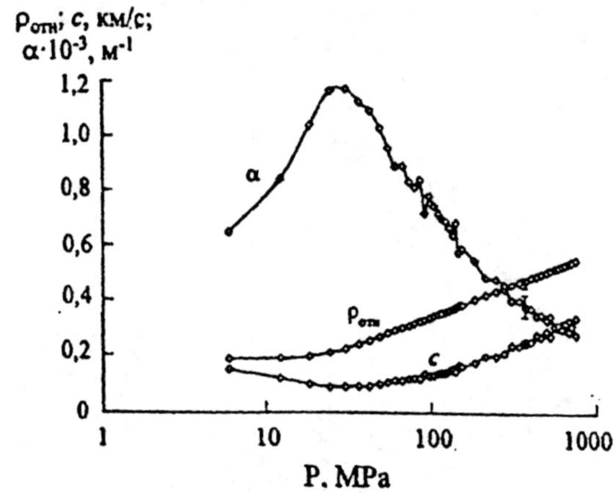

Figure 4 The dependencies of sound speed $c(P)$, attenuation of ultrasound $\alpha(P)$, the densification curve of SDP ZrO_2-5.4%Y_2O_3 in the case of US-pressing (W=1 kW).

Discussing Results

Let us note the following peculiarities of the densification curves (Fig.1):

The value of P_c for the flex points of densification curves have the tendency to the increase with power of USV, excepting W=0.25 kW (Fig.2, Table 1).

With increasing the power of USV the tilting angle of densification curve on the beginning stage of the pressing decreases, and for values W=1.5-3 kW in the mentioned region the curves are practically parallel to the pressure axis.

Despite of identical beginning conditions of the pressing of SDP in all the experiments, the current density of powder body on the beginning stage of the pressing exhibits the dependence vs. USV power level (the densification curve family 1-7 is "split" in the region of small P proportionally to value W, excepting the minimal level W=0.25 kW).

The observed behavior of densification curves can be explained when to assume that energy of US-exposure makes the energy barrier for compressing the bulk powder. On this stage of the pressing the SDP is the quasi-elastic gas-dispersed medium characterized by the corresponding value of uniform compression module of $K(P)$ – Fig.3, the speed and sound absorption $c(P)$, $\alpha(P)$ – Fig.4. Under the static condition the initial SDP has certain friction coefficient, determined by equation of the statics of the loose medium (Coulomb's law) [3]. The SDP excites the vibration shifts ξ of powder particles and the oscillation component of elastic bonds between particles of SDP appears, which is the cohesion rigidity of SDP particles k:

$$k=m\omega_0^2, \qquad (18)$$

where ω_0 is proper frequency of SDP particle oscillations of mass m in US-field.

Thereby the US-exposure definitely generates the structure in the powder body, transferring it from the medium consisted of discrete particles into the vibrating continuum, having a higher module of bulk modulus of elasticity (Fig.3). The volume of such a vibrating continuum is inversely proportional to the sound attenuation constant in the powder α. For overcoming the elasticity of the vibrating powder a higher values of the pressure $P>P_c$ are required (Fig.1,2; Table 1). Therefore until the P_c is not achieved the density of the vibrating powder weakly depends on the P (beginning parts of curves 3-7 in Fig.1), in contrast to static pressing (curve 1).

On the beginning stages of the pressing the conditions of passing the ultrasonic through loose powder body are essentially changed: the freely raised powder behaves itself as the gas-dispersed medium, in which the USV are effectively attenuated. After the interparticle bonds in powder at $P=P_c$ have become a stronger, there are conditions for passing the sound through the pressed billet in both the solid and the porous body. Such a conclusion is confirmed by nonmonotomic behavior of the speed and the absorption of USV in Fig.4.

The influence of USV exposure on the density of the pressed billets is essential. After USV having been feed to the mold, when the static pressure was not supplied, the shift of ram was observed, which increases with the power W that is reflected in the significant increase of the pressed body density, being proportional to the USV power (the splitting of the beginning parts of the densification curves in Fig.1). The mentioned fact is explained by increasing the interparticle and wall friction in the powder, which is in the field of US-vibrations [4,8].

On the beginning stage of the pressing at the USV of small intensity (W=0.25 kW) a less density of SDP comparatively to static pressure was observed. In this case due to large attenuation of sound of small intensity only the thin external layer of USV was undergone to US-exposure from the matrix walls. Thus the decrease of the interparticle friction in bulk of the pressed billet is not significant, the increase of elasticity of sonicated external layer resists to the compression of powder in the region of small P [8].

However, the final density of SDP pressed billet and especially of the ceramics sintered of it, does not unambiguously correlate with the value W (the incut in Fig.1 and Table 1): the maximal density of the pressed billets of SDP is ρ=3.39 g/cm^3 (56% of theoretical density of tetragonal modification zirconium ceramics) was achieved in the cases W=1.5 and 3 kW. But, in the ceramics, the maximal density is 5.63 g/cm^3, which corresponds to the samples, pressed at W=1 kW.

The given data demonstrate that US-exposure during the pressing essentially influences on the ceramics density, but the maximal density of the pressed billet does not provides the maximal density of the ceramics. The density uniformity of the pressed body is obvious to be the determinative factor. Depending on the power W, influencing on the appearance of the acoustic streams in SDP, the ultrasound can both to homogenize the pressed billet and to create the density gradients in it. Thus, the intensity (power density) of USV is of important significance (8), which can be larger or lesser of the critical value I_c, defined to (13).

In Fig.4 the dependence $I_c(P)$ for the pressed SDP is presented.

For $P<P_c$ in the case of US-exposure of high intensity $I>I_c$ (region 1 in Fig.5) the powder body effectively absorbs the ultrasound (Fig.4), the dissipation of sound energy leads to the disruption of agglomerates, to the increase of powder dispersity with changing the particle morphology and by their mechanical activation. The mass transfer of SDP particles, occurring due to the appearance of acoustic streams, favours to this as well. Under the USV-exposure of low intensity $I<I_c$ (region II in Fig.5) the acoustic stream is absent, and when decreasing the forces of interparticle and the wall friction, the SDP particles are uniformly distributed in the pressed billet.

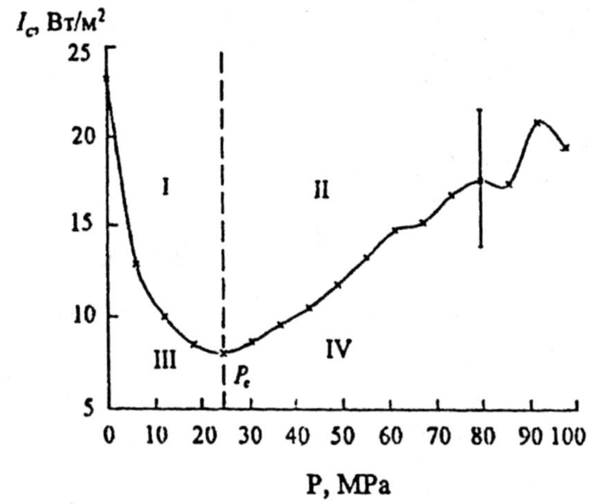

Figure 5 The critical intensity of USV $I_c(P)$ in the SDP ZrO_2-5.4%Y_2O_3.

After high-power US-exposure over SDP during the further pressure ($P>P_c$, $I>I_c$, the region II in Fig.5) the sharper decrease of elastic aftereffect δ from 0.40% to 0.27% was observed (Table 1), that is explained by the significant change of elastic-plastic properties of SDP particles under high-power ultrasonic exposure. In a given case the appearance of effect Blag-Langenekker is possible, when the US-exposure upon the statically deformed material leads to increasing the plasticity of the latter [6]. However, for the relatively large amplitudes of the vibration shifts of SDP particles $|\xi|\geq d$, the density distribution in the bulk of the pressed billet becomes non-uniform, that is reflected in the decrease of the density of the sintered ceramics (Table 1).

For $P>P_c$ the ultrasonic absorption in powder body occurs to the less degree due to the increasing of powder friction constant. The dissipation of ultrasonic energy for $|\xi|\geq d$ ($I>I_c$ region IV in Fig.5) favours to the activation of the particle contact surface; at the same time the powder is uniformly densified. The vibrations of particles of SDP under sonic pressure p resists to their coagulation even in the dense pressed billet, preventing the plastic deformation of particles at the final stage of the pressure. The pointed effects favours to preserving the nanoparticles of the powder in dense pressed bodies as the crystallization nucleuses for further ceramics sintering. The grain of zirconium ceramics, sintered from statically pressed SDP, was 0.8 μm, in the case of US-pressing – 0.3 μm, when the subgrain structure (texture) of nanoceramics of layer thickness from 0.09 up to 0.1 μm in the grain was observed as well [10].

Issuing from data of Table 1, the generator power of 1 kW W_c 1.5 kW was corresponding to value I_c in a performed experiments.

In Table 2, the parameters *c*, *p*, *z*, α, I_c, calculated to equations (1)-(10), (13) for superdispersed (*d*=10 μm) and large-dispersed (*d*=10 μm) powders of zirconium dioxide are compared: for the freely raised (*P*=0) and at *P*=740 MPa.

It follows from Table 2 that for dense pressing of SDP the value I_c is essentially smaller than for large-dispersed powder, despite of that conditions of the ultrasonic propagation in SDP are worse (the speed *c* is smaller, the sound attenuation α is larger than in the large-dispersed powder). This means that for dense pressing of SDP the small intensity of US-exposure is enough. This conclusion is confirmed by the densification curves (the incut in Fig.1, Table 1); already at US-exposure upon the SDP from *W*=0.25 kW the density of the compact increases in the comparison to the case of static pressing.

Table 1 The influence of USV power on the parameters of the pressed billets of SDP (ZrO_2-5.4%Y_2O_3) and sintered ceramics

Power of USV *W*, kW	0 (static pressing)	0.25	0.5	1	1.5	2	3
Critical pressing pressure of SDP P_c, ±3	24	20	25	30	30	31	37
MPa	3.26	3.28	3.34	3.35	3.39	3.36	3.39
Pressed billet density, ±0.07 g/cm^3	5.23	5341	5.58	5.63	5.60	5.35	5.32
Density of sintered ceramics, ±0.07 g/cm^3	0.57	0.57	0.52	0.45	0.40	0.27	0.25
Elastic aftereffect δ, %							

Table 2 Parameters of US-exposure on the powder (ZrO_2-Y_2O_3) of different dispersity

Parameter	Freely raised powder		The pressure *P*=740 MPa	
	d=0.1μm	d=10μm	d=0.1μm	d=10μm
c, m/s	265	321	338	423
p, kPa	3.69	1.91	14.21	8.66
z, MPa·s/m	0.294	0.152	1.13	0.688
α, m^{-1}	355.7	117.3	279.1	89.1
I_c, kW/m^2	0.023	1.2	0.089	5.44

CONCLUSION

During the US-pressing of SDP the speed, the intensity, the sound attenuation, and other acoustic parameters connected with them are changing in nonmonotone way when changing the powder density. On the beginning stages of the pressing the freely raised powder behaves oneself as the gas-dispersed medium, and after the setting of a more strong interparticle bonds in the powder at $P=P_c$ the conditions for propagation of sound through the pressed billet are similar to ones of porous body.

The obtaining of the maximal dense pressed billets of SDP occurs when neither the vibration shift of the particles does not exceed their sizes. This condition determines the critical value of US-intensity I_c above which in the powder at the beginning stages of the pressing ($P<P_c$) the acoustic streams can occur, and further ($P>P_c$) the elastic-plastic properties of SDP particles significantly change.

Depending on the ratio of the regimes of the pressing *P*, *I* with critical parameters P_c and I_c, four types of US-exposure upon SDP are possible:

$P<Pc$, $I>I_c$ is the disruption of the agglomerates, the activation of SDP particles under conditions of acoustic streams;

$P<Pc$, $I<I_c$ is uniform densification of SDP at the beginning stages of the pressing;

$P>Pc$, $I>I_c$ is the pressing of SDP with changing the elastic-plastic properties of powder;

$P>Pc$, $I<I_c$ is uniform-density pressing of SDP with preserving the nanostructure of the pressed body.

The work is particularly supported by grants of Ministry of secondary and professional education of RF to the division "Powder Metallurgy" (# 97-18-3.3-78) and the division "Fundamental and theoretical bases of new resources-economy technologies".

REFERENCES

1. Shevchenko V.Ya., Barinov S.M. Tehnicheskaya keramika. M., Nauka, 1993, 197s.
2. Gusev A.I. Nanokristallicheskie materialy: metody polucheniya i svoystva. Ekaterinburg, UrO RAN, 1998, 199s.
3. Andrievskiy R.A. Poroshkovoe metarialovedenie. M., Metallurgiya, 1991, 205 s.
4. Agranat B.A., Gudovich A.P., Nezhevenko L.B. Ultra-zvuk v poroshkovoy metallurgii. M., Metallurgiya, 1986,168 s.
5. Roman O.V., Skorohod V.V., Fridman G.R. Ultra-zvukovoy i rezistometricheskiy kontrol v po-roshkovoy metallurgii. Minsk, Vysheyshaya shkola, 1989,182 s.
6. Ultrazvuk. Malenkaya entsoklopediya. Gl. red. I.P. Golyamina. M., Sovetskaya entsiklopediya, 1979, 400 s.
7. Babikov O.I. Ultrazvuk i ego primenenie v pro-myshlennosti. M., Fizmatgiz, 1958, 260 s.
8. Khasanov O.L., Poholkov Yu.P., Sokolov V.M. i dr. Steklo i keramika, 1995, no.7, s.15-18
9. Glass S.J. and Ewsuk K.G. MRS Bulletin, 1997, V.22, No.12,p.24-28.
10. Khasanov O.L., Poholkov Yu.P., Sokolov V.M. i dr. V sb.: Fiziko-himiya ultradispersnyh sistem (Materialy IV Vserossiyskoy konferentsii). M., MIFI, 1998,c.280-282.

ABOUT THE AUTHORS

Khasanov Oleg Leonidovich *– Tomsk Polytechnic University. A Candidate of Physics and Mathematics, director of research center "Spektr" of TPU. An expert in area of materials science of superdispersed materials.*

Dvilis Edgar Sergeevich *– Tomsk Polytechnic University. A postgraduate student. An expert in area of ultrasonic treatment of materials.*

Pokholkov Yuri Petrovich *– Tomsk Polytechnic University. A rector, doctor of technical sciences. An expert in area of materials science of electrotechnical materials.*

Sokolov Vitaliy Mikhaylovich *– Tomsk Polytechnic University. A candidate of technical sciences, research officer of centers "Spektr" TPU. An expert in area of ultrasonic technology.*

USING CAPACITY METHOD TO STUDY DIELECTRIC LAYERS HETEROGENEITY

V.P.Pronin and S.A.Pet'kov

Using multiparameter capacity system is expedient for investigating heterogeneity of dielectric layers. With the help of the developed devices the experiments connected with peculiarities of electrization of section-homogeneous layers in the field of corona discharge on the boundary of two dielectrics with different dielectric permittivity and conductivity have been carried out. The obtained results are in agreement with general idea of dielectric electrization. The experimental investigations on electrization of the dielectric surface layer with arbitrary relief are the evidences of wide functional possibilities of the capacity method and the developed devices.

For wrap-around measuring the distribution of dielectric materials electrical properties in solving a lot of problems of dielectric physics the use of capacity system is prospective. The simultaneous control of several parameters of material is often set in the practice. This is especially important during investigation of dielectric layer homogeneity, therefore the use of multi-parameter capacity systems is the expedient.

The electrization of homogenous layer depends on several parameters – the distance, the voltage, the corona exposure time, the thickness, the conductivity and the layer permittivity, and besides, on their distribution along the layer. In this connection, during the investigation of inhomogeneous layers, it is necessary to know how the electrization time influences upon the defect revealing [1].

For solving this problem with the help of the developed devices realizing the multi-parameter investigations, the experiments of electrization of the layer of thickness of H=0.5 mm, consisting of the set of different dielectrics (Fig.1), have been carried out. The electrization time was changed from hundredth parts to the several seconds, the sample movement velocity is 10 m/s, the diameter of electrode is 0.5 mm, and the minimal width of dielectrics is 2 mm.

It follows from the oscillograms, when the sensing electrode is in the central region of every dielectric the charge distribution is supposed to be constant, therefore, assuming the transfer coefficient is constant component [2], the charge density on the n-th dielectric is defined as

$$\sigma_n = \frac{U_{0n} C_{BX}}{K} \frac{H + \varepsilon_n h}{HS}$$

the ration of the charge density on n-th and i-th dielectrics is

$$\frac{\sigma_n}{\sigma_i} = \frac{U_{0n}}{U_{0i}} \frac{H + \varepsilon_n h}{H + \varepsilon_i h}$$

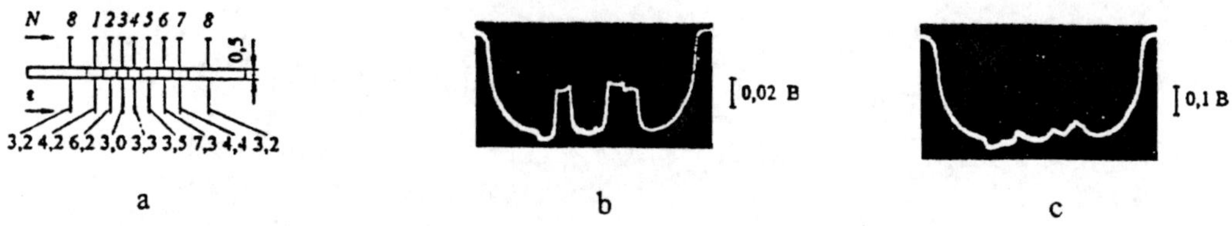

Figure 1. *a* – the layer model; *b,c* – the responses of sensing electrode system on the distribution of charge for time of electrization of 10^{-2} s and 3 s, respectively: 1 – the epoxy resin, 2 – the getinaks, 3 – the plastic, 4,8 – the plexiglass, 5 – the ebonite, 6 – the textolite, 7 – the carton with epoxy treatment.

For a certain H=0.5 mm and h=0.1 mm, and for a certain values of permittivity of every dielectric, one can find this ratio at small (0.01-0.1 s) time of layer electrization.

Figure 2. Appearance of studied relief of dielectric surface: 1 – the periodic relief with the constant hollow depth; 2 – the relief with variable hollow depth δ=0.1; 0.2; 0.3; 0.4; 0.5; 0.6; 0.7 mm; 3 – the relief with the varied Δd=0.10; 0.15; 0.20; 0.30; 0.40; 0.50; 0.70; 1.0 mm and δ=const; 4 – the relief with varied δ, Δd, α.

For determining the permittivity the following technique is used. In the system the quasi-static electrical field which is transversal to the layer is applied. In the sensing electrode circuit both the induced charge at the frequencies defined by the scanning speed and the charge caused by the change of permittivity along the layer at constant H and h are measured [3]. The permittivity is given in Fig.1. Having used them, choosing, for example, the plexiglass as the i-th layer, we obtain:

$$\frac{\sigma_n}{\sigma_i} = 1.29;\ 1.21;\ 1.26;\ 1.20;\ 1.31;\ 1.03;\ 0.74;$$

n= 1,2, …

The ratios of charge densities on the parts of layer with the different permittivity are little differed each from other (in order of 10%). The significant difference σ_6/σ_8 and σ_7/σ_8 from the previous ratios is explained by large conductivity enough of textolite and carton, leading to the fast discharge of these layer parts.

Under long-term electrization of section-homogenous layer the charge distribution is essentially heterogeneous, the potential distribution insignificantly differs from average level.

All this agrees with general idea about electrization of dielectric [4].

With the help of the developed devices, one can experimentally ascertain what the mean in the field of corona discharge the surface of dielectric layer with arbitrary relief is electrified and how the charge after electrization is distributed. These processes are practically unstudied but are of large significance in different area of physics and technics [5,6]. The analytical solutions of these problems are difficult, but the carried out experimental investigations allow to make some conclusions with regard to the response of sensing electrode system on the charge distribution. In particular, four kinds of one-dimension relief of the surface exhibited in Fig.2 have been investigated.

The oscillograms of responses of sensing electrode system with the diameter of the signal electrode of 0.3 mm upon the charge distribution are given in Fig.3.

During the electrization of homogenous layers it has been determined that charge on the surface is in inverse proportion to the layer thickness. Consequently, when the sensing electrode is between the parts, which are in one plane parallel to the scanning plane, this charge can be equal. The obtained experimental measure of the signal during the scanning along the surface, in general, replicates its relief, however, for explanation of the observed quantitative relation between signal value and hollow depth it is necessary in the future to investigate the behaviors of accumulation and relaxation of the charge on the relief surface during and after its treatment by corona discharge.

CONCLUSION

1. The capacity method of probing the charge distribution, settled on the dielectric surface from the plasma of corona discharge allows:
   - to determine the location of macro-heterogeneity (of size not less than 0.3 mm), engaged into dielectric bulk;
   - to determine the relief of dielectric surface and to reveal the places differing each from other with the thickness. It should be noted that in the case of deep hollows over the dielectric surface the value of the measured signal is little different from the expected one and the carrying out of additional investigations is necessary.
2. The results of investigations have shown that technique described in the work can be used for controlling the electrical and physical parameters and the quality of dielectric production

REFERENCES

1. Grischenko V.L., Matveeva I.A. Preobrazovanie poverhnostnyh raspredeleniy zaryadov i potentsiala metodom elektrostaticheskoy induktsii. V kn. Voprosy elektroniki SVCh. Nekotorye problemy radiofiziki. SGU,1983,s.16-24.
2. Matveeva I.A. Radiotehnika i elektronika, 1982,no.4, e.804-812
3. Uorsing A. Metody obrabotki eksperimentalnyh dannyh. M.: Inostrannaya literatura, 1953, 348 s.
4. Skanavi G.I. Fizika dielektrikov (oblast slabyh poley). M.: Gostehizdat, 1949, 500 s.
5. Gubkin A.N. Elektrety. M.: Nauka, 1978, 172 s.
6. Guscho Yu.P. Fazovaya relefografiya. M.: Energiya, 1974,137 s.

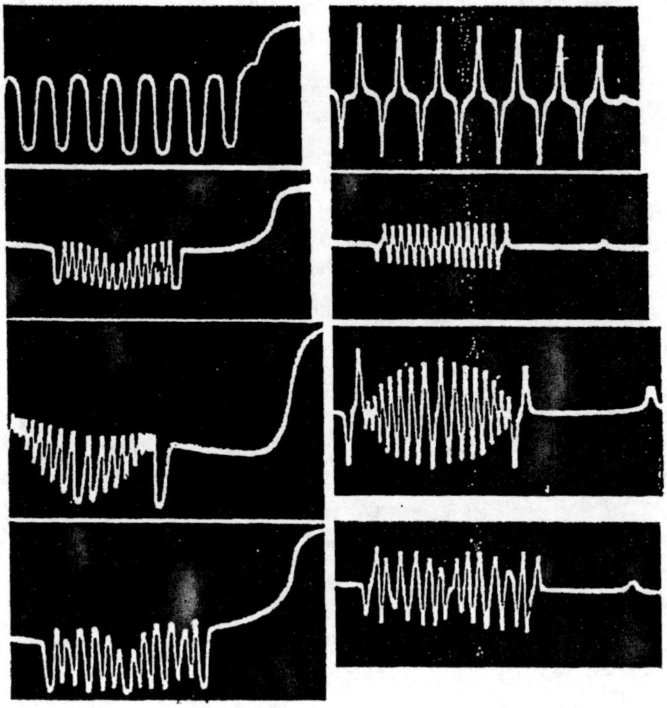

a b

Figure 3. The responses of sensing electrode system with the line resolution of 0.5 mm in the charge (*a*) and gradient (*b*) indication mode at electrization of relief dielectric surfaces 1, 2, 3, 4 in the field of negative corona discharge.

About the Authors

Pronin Vitaliy Petrovich – *Saratov State Agrarian University after N.I.Vavilov. A professor, doctor of Technical sciences, head of chair. An expert in area of dielectric physics.*

Pet'kov Sergey Aleksadrovich – *Saratov State Agrarian University after N.I.Vavilov. A postgraduate student. An expert in area of solid state physics, development of mathematical support of check technique of electrical and physical parameters and quality of dielectric production.*

MULTIFRACTAL ANALYSIS OF LASER-TREATED COPPER STRUCTURE

A.G. Kolmakov, G.V. Vstovskiy, S.A. Maslyaev and V.N. Pimenov

By using the multifractal concept the changes of grain structure in the plates of copper alloys Cu-4%Ni, Cu-2%Al (mass.%) of thickness of 0.2 mm in zone of through penetration with laser pulse of wave length of 1.06 μm in the free generation regime have been studied. The layer epitaxial growth crystallization had led to the formation of specific "petalous" grain structure with predomination of strongly elongated grains. The use of multifractal concepts had revealed that the changes in the grain structures for all the studied alloys, caused by laser treatment, in general, are of the equal character and are followed with changes of the same kind of multifractal characteristics – the decrease of homogeneity and the latent orderliness in the zone of "laser hardening". Such a character of structural changes allows assuming the decrease of mechanical properties of materials in the zone of laser treatment. In this respect, the alloy Cu-2%Al is the most sensitive one. The multifractal parameterization of the structures has shown one's effectiveness as the instrument for the image identification, in particular, for identification of the grain structures and for revelation the raw of the unrevealed behaviors when using the traditional approaches.

INTRODUCTION

The use of laser emission (LE) as the instrument for treating the materials for the aim of improving their properties is supposed to be one of the prospective directions in the area of development of the new materials and technologies [1,2]. The study of fault probability of materials and the changes of their structure and properties, subjected to non-stationary heat streams like the plasma disruption, caused by high-power laser pulses of LE, is of great interest [3-5]. This question, besides of itself interest, is of great interest for the problem of development of materials for thermonuclear synthesis reactors [6]. In this connection, the investigations of effect pulse LE upon the grain structure of metallic materials, and, consequently their mechanical properties is possessing special actuality.

The description of grain structures is traditionally provided to the quantitative parameters characterizing their separate elements in combination with data statistic treatment (the grain average size, the grain size distribution, the density of the grain boundaries etc). There are also known the results of works (their overview is given in [7]), which attempt to describe the peculiarities of grain structures (by other words – the grain boundaries configurations) in metallic alloys with using the fractal theory methods. Bringing the multifractal conception, based on using the general notion of the measure, allows giving the quantitative evaluation of configuration of the studied structure in general, that supplies the conventional methods of quantitative metallography.

In particular, the technique of multifractal parameterization of the structures, developed on the base of original information-theoretical interpretation of multifractal formalism (MF) [8], allows introducing and investigating the quantitative characteristics of homogeneity and the order of the structures [9-14].

In present work, by using the multifractal parameterization technique, the changes of grain structure in laser-treated copper alloys thin plates have been studied.

MATERIALS AND TECHNIQUES

The plates of thickness of 0.2 mm after the rolling and recrystallization annealing (T=0.5 T_{ml}) of copper alloys Cu-4%Ni, Cu-10%Ga, Cu-2% Al (mass.%) were used as the experimental specimens. The specimens were treated with the LE, the through penetration zone had expanded to the whole depth of the plate without interrupting the liquid phase [15,16]. The specimens were treated in the air by means of the plant GOS-1001 of LE length of 1.06 μm in free generation regime of pulse duration of $0.7 \cdot 10^{-3}$ s for the intensity of emission of the order of $5 \cdot 10^5$ W/cm^2 and of diameter of the beam of 2-3 mm [15].

The grain structures were studied to the light microscopy. The maximum and average size of the grains in the direction of their oblongness and the minimum size of grains were determined. The average size of grains in the direction of their oblongness was determined from measurement of sizes of 35 grains on the micro-pictures of the thin sections. For describing the peculiarities of configurations of the studied structures, the original technique of multifractal parameterization of the structures of materials [9-10] realized to computer software, have been used. In short, the essence of the technique reduces to the performing of the three stages.

1. The identification of the grain structures: the fragmentation of microstructure pictures with the square net, consisting of 256×256 equal cells. The cells related to the boundaries of grains are assigned to "1", the cells related to the body of the grains are assigned to"0". Thus the grain structures were approached to the digit.
2. The input of the obtained arrays of the digit values (the matrixes or bit maps, in which the minimal unit of information "bit" has two states: "1" or "0") into the cells of computer memory for the following mathematical treatment (correlation analysis).
3. The fragmentation of the matrix onto more large cells with the dimensions of $l_k \times l_k$ (k=1,...,6, $l_k=2^{(k+1)}$=4, 8, 16, 32,64, 128) and the building of the characteristic measure for every fragmentation in the form of equal-cell distribution of units P_i ($P_i=M_i/\Sigma M_i$, where M_i is the number of the units in i-th large cell, ΣM_i is a total number of units in the matrix of large cells, i=1, 2, 3,...,N; $N=[256/l_k]^2$); the next step is the calculation of the conventional MF characteristics such as $f(\alpha)$-spectra and D_q-spectra of the Renii's dimension [17,18] for the set of values q from the interval [-40;40]. The technique allows quantitatively evaluating the relative homogeneity (f_{40}) and the latent orderliness ($K=D_{-40}$-D_{40} or $\Delta_{40}=D_1$-D_{40}) of the structures. The larger f_{40} the more homogenous structure; the larger K [19] or Δ_{40} the more structure is ordered [8-14]. Due to virtue of the reasons of the purely methodological character, the use of the parameter Δ_{40} is more appropriative [9-11], therefore this parameter was used for evaluating the periodicity. While considering the multifractal characteristic f_{40}, it should be noted that hence the homogeneity implies not the conventional quality characteristic of the structure appearance but the character of the distribution of unit elements of the considered

structure (one cell ~ 1 pixel~1 bit of information) in the Euclidean space. In a given case, we mean the elements corresponding to the grain boundaries in the studied square area of thin section plane (approximated images).

Since for the multifractal analysis there are considered the distribution of any value on the area, then in a given case the distribution of elements in a detailed square area of the plane of thin section, corresponding to the grain boundaries, were studied. The dimension of the studied areas of structure was 350×350 μm, the dimension of one cell is equal to 1.37×1.37 μm. Such the dimension of the areas exceeds the thickness of source plates and has the near extent as the diameter of the melted zone. In this connection, the chosen dimension of studied areas was correct enough from both the point of view of conventional approaches and the applicability of multifractal parameterization.

Discussing the Results

The approximated pictures of areas of grain structures of copper alloys before and after laser treatment are given in Fig.1. The data on sizes of grains and some calculated multifractal characteristics for studied structures are given in table. The portion (in percentage) of elemental cells related to the grain boundaries in approximated pictures of grain structures is presented there as well. The obtained pictures of D_q-spectra of Renii's measures and $f(\alpha)$-spectra are shown, respectively, in Fig.2 and Fig. 3. The change of latent order and homogeneity of grain structures of copper alloys, resulting from LE treatment, is shown in Fig.4.

Analysis of Grain Structures in Initial State and After Laser Treatment

The structure of copper alloys in initial state (Fig.1.1) consists of grains slightly elongated to the direction of preceding rolling. The grains are of the triangle, rectangle and polyhedron shape. Among the alloys Cu-4%Ni, Cu-10%Ga and Cu-2%Al, the sizes of grains increase, the density of grain structure (number of grains on the studied square area) decreases (Table). The dispersion of grain sizes is not large, excepting the alloy Cu-10%Ga. In structure of alloys Cu-4%Ni and Cu-2%Al, the sharp separation of grains onto groups for the size and shape is practically absent. In the structure of alloys Cu-10%Ga the grains can be conditionally separated onto three groups: the weakly-elongated grains with respectively large and small sizes and strongly-elongated grains of rectangular shape. For alloy Cu-2%Al, the boundaries of grains are smooth enough, the grains of alloys Cu-4%Ni and Cu-10%Ga can be characterized as more indented ones.

The LE treatment has caused the melting of alloys first, and after, the liquid phase crystallization. The hardening passed from the periphery of the area subjected to laser treatment to its center. For all the alloys there is observed the significant increase of sizes of the grains. The shape of grains have significantly changed: the grains are strongly elongated, predominantly along the crystallization direction, forming the specific "petalous" structure (Fig. 1.2). The density of grain structure (number of grains on the studying square area) decreases, the sizes of grains increase in the alloys Cu-4%Ni, Cu-10%Ga and Cu-2%Al, i.e. in the manner as for initials state of structure. For alloys Cu-4%Ni and Cu-2%Al the grain size dispersion strongly increased (about for extent). For alloy Cu-10%Ga the increase of the grain size dispersion is less expressed. In structure of alloy Cu-2%Al there is the significant angle random orientation in the direction of the grain elongation; for the alloy Cu-10%Ga the most grains are oriented in one direction. For alloys Cu-2%Al and Cu-4%Ni the boundaries of grains, in general, are smooth, and are indented for alloy Cu-10%Ga.

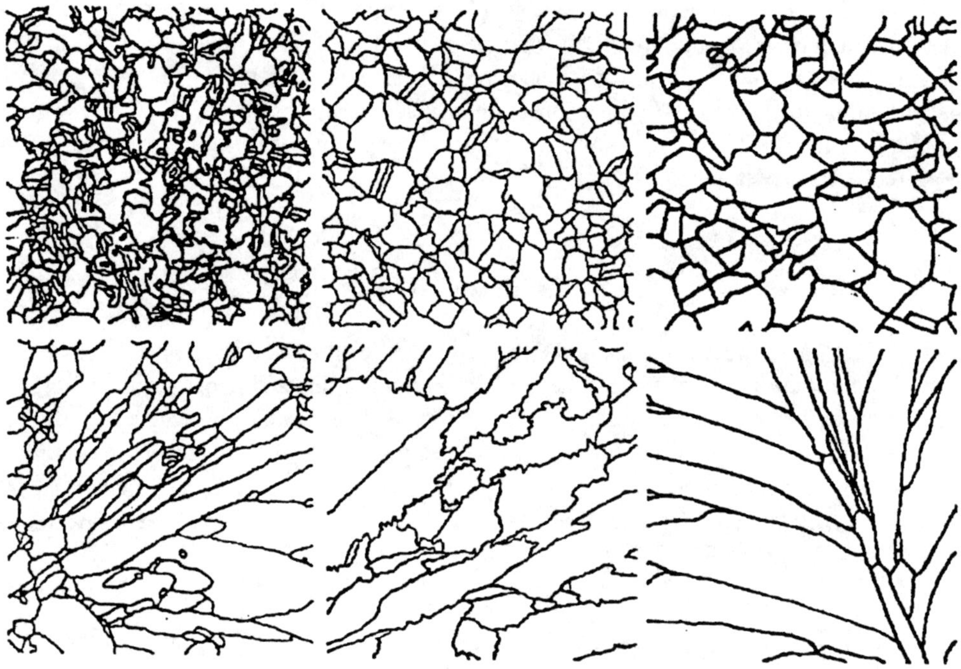

1 2 a b c

Figure 1. The approximated images of parts of copper alloy grain structure of dimension of 350×350 μm: 1 – the initial structures, 2 – after laser emission treatment; a – Cu-4%Ni, b – Cu-10%Ga, c – Cu-2%Al.

It should be noted that formation of the "petalous" structure under laser treatment, to all appearances, is typical exactly for thin specimens. For massive specimens, the LE treatment being analogous in the regimes, had caused the formation of the grain structure of absolutely other type – fine-grained dendrite, super dispersed and even the quasi-amorphous one [1,2,20]. It is explained by that cooling rate of the zone melted by laser emission in first case was approximately 10^3 K/s, that is for two-three degree lower than in the second case [15].

Early in works [15,16] it has been demonstrated that chemical elements distribution over the grain was homogeneous and did not change after laser emission treatment for all the studied alloys, the significant macro-segregations of the dopants were absent; the texture after laser treatment coincided with initial structure and became more expressed as well. Taking into consideration all these data, one can expect that grain structure formation mechanism was of epitaxial character. It coincides with conclusions made in [20] about epitaxial growth of grains under laser fusion of surface.

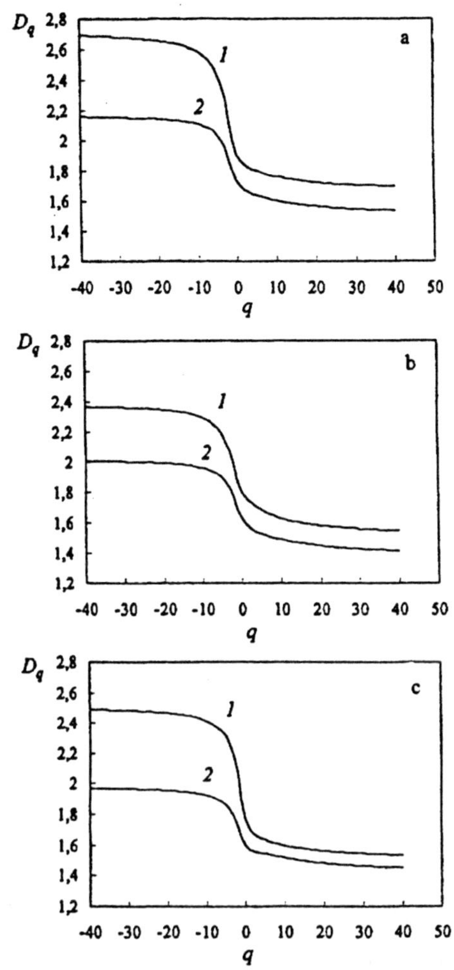

Figure 2. The diagrams of D_q-spectra of Renii's dimensions, obtained for studied parts of grain structures of copper alloys:
1 – the initial structures, 2 – after laser emission treatment;
a – Cu-4%Ni, b- Cu-10%Ga, c – Cu-2%Al.

Analysis of Multifractal Characteristics of Studied Structures

The multifractal spectra for initial grain structures and structures appeared as the result of laser through penetration are significantly different; the spectra shift for all investigated alloys are of the same character (Fig.2 and 3, Table). While index $q \geq 1$ increases, the value of shift of the generalized Renii's dimensions D_q and indices α_q (q=0-40) stabilize. It is explained by that with increasing $q>1$, the inaccuracy of determination of characteristic measure P_i is negligibly small even for essential experimental and calculated inaccuracies, and the sensitivity to the properties of distribution $\{P_i\}$ increases [9-11]. This fact is the evidence of limitation of the fractal parameterization of structures (only for the value of fractal dimension D_0), in comparison to multifractal parameterization. The relative shift of all Renii's dimensions (at $q \geq 0$) for studied alloys is equal about. The relation of the values of characteristics for the structures before and after laser treatment is: 1.1 for Cu-4%Ni and Cu-10%Ga, 1.06 for Cu-2%Al. The insignificant different for latter alloy can be explained, first of all, by small amount of Al, and by those Ni and Ga being placed in periodic table in one period as Cu; Al is placed in neighboring one (because of difference of building of the complete internal electron shells [21]). Such a result is quite natural. Theoretically, multifractal characteristics in somehow correspond to thermodynamic features of system, on which the measure distribution is studied. Although there was not unambiguously compared to concrete thermodynamic characteristics, but in every case for particular models a certain multifractal characteristics were interpreted as the temperature, entropy, free energy [8,22-29]. In particular, the generalized Renii's dimensions corresponding to the Renii's entropies were introduced early as one of the possible presentation of the statistical sum [17]. In this connection, the considered multifractal characteristics of grain structures must have some quantitative information of thermodynamic conditions of their obtaining. As the initial grain structures of studied copper alloys were produced under the same technological conditions and the following LE treatment was also carried out under practically the same conditions, the above mentioned relative value of the shift for them must be identical.

In initial state the grain structure of alloy Cu-4%Ni is of the greatest homogeneity, the alloy Cu-10%Ga has the smallest one. In first case the larger homogeneity of distribution of elements corresponding to the grain boundaries is caused by fine-grain structure of high density, by small dispersion of sizes of elements and by the absence of sharp separation of grain onto the groups in the sizes and shapes. In second case the small homogeneity is connected with very strong grain size dispersion and with the outlining separation of grains onto the groups in sizes and shapes. After laser treatment the structures of the same alloys were respectively distinguished with the largest and smallest homogeneity. For alloy Cu-4%Ni it is connected with that, despite of sizes grain dispersion is great enough, their separation onto the groups and the irregularity of the grain boundaries is practically absent, the structure is characterized with the smallest grain sizes and with the greatest density. For alloy Cu-10%Ga the small values f_{40} are connected rather with strong irregularity of the grain boundaries.

The grain structure of alloy Cu-10%Ga is distinguished with the greatest orderliness, for initial state. That, obviously, is mainly connected with formation in the grain structures of three sub-structures (sub-systems) which enlarge the order. The structures of the alloys Cu-4%Ni and Cu-2%Al, which have not such a separation, are distinguished with less and very near values of Δ_{40}. After laser treatment, the structure of alloy Cu-10%Ga that is characterized with the smallest sizes grain dispersion and with the domination of large and middle grain having practically equal elongation direction, has the largest value Δ_{40} as well. The grain structure of alloy Cu-2%Al in comparison to other alloys is distinguished with the largest disordered orientation, with the largest sizes of grains and with the smallest density. In this connection the portion of chaotic component in the structure of this alloy is the largest and, respectively, the value Δ_{40} is the smallest.

The multifractal characteristics, reflecting the homogeneity (f_{40}) and latent orderliness (Δ_{40}) of grain structures, are of integral character and reflect all the peculiarities of every studied structure and not only peculiarities of its obtaining. In this connection the described above for D_q and α_q the quality dependence of the shift for the transition from initial structures to the laser-treated structures for these characteristics is absence (Table). Nevertheless the general quality character of change of characteristics f_{40} and Δ_{40} is equal for all investigated alloys: the homogeneity and the degree of latent orderliness of grain structures appeared after laser treatment decreases in comparison to their values for initial structures (the arrows in Fig.4).

Table: The Grain Sizes and Some Calculated Multifractal Characteristics for Studied Structures of Copper Alloys Before and After Laser Exposure (LT)

Alloy	Cu-4%Ni		Cu-2%Al		Cu-10%Ga	
Structure	initial	After LT	initial	After LT	initial	After LT
	The sizes of grains					
Maximum in the direction of the grain oblongness, μm	37	335	125	640	60	370
An average to the direction of grain oblongness, μm	13	72	45	280	25	50
Minimal, μm	3	3	10	6	1	5
The relation of maximum and minimal sizes	12	112	13	108	60	74
The density of grain structure (a number of grains on the studied parts)	517	150	129	36	226	59
The relative decrease of density of grain structure	3.4		3.6		3.8	
The portion of elementary cells, which are on the grain boundaries in approximated images of grain structures						
%	26.3	9.2	18.2	9.2	12.5	8.0
	Multifractal characteristics					
D_0	1.191416	1.73517	1.7728	1.5964	1.80805	1.63549
$D_1=f_1=\alpha_1$	1.187146	1.70018	1.171158	1.56993	1.76717	1.59596
D_2	1.84632	1.6787	1.68121	1.56015	1.73991	1.57085
D_{40}	1.69763	1.53698	1.53633	1.45496	1.54856	1.41407
α_{40}	1.66816	1.5051	1.50688	1.42194	1.5413	1.38036
f_{40}	0.51899	0.26183	0.35856	0.13439	0.17797	0.06592
$\Delta_{40}=D_1-D_{40}$	0.17383	0.1632	0.17525	0.11497	0.21861	0.18189
The relative decrease of f_{40}	1.98		2.68		2.7	
The relative decrease of f_{40}	1.07		1.52		1.2	

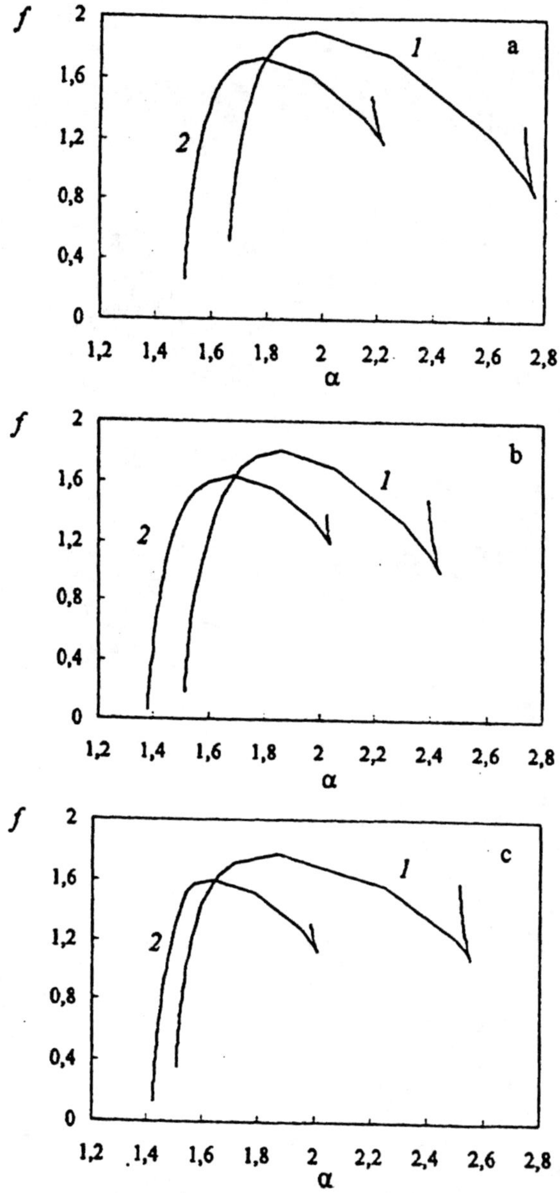

Figure 3. The diagrams of f(α)-spectra, obtained for the studied parts of grain structures of copper alloys: 1 – initial structures, 2 – after laser treatment; a – Cu-4%Ni, b – Cu-10%Ga, c – Cu-2%Al.

The decrease of the homogeneity index f_{40} of structures is mostly connected with change of the irregularity of boundaries and particularly with increasing the grain sizes in structure and with decreasing its density (the decrease of general portion of elements of structure, corresponding to grain boundaries). The essential relative decrease of homogeneity was observed for alloy Cu-10%Ga and the smallest – for alloy Cu-4%Ni, since in first case, besides of the increase of sizes of grains, the appearance of irregularity of their boundaries and the largest relative decrease of density of grain structure occurs, in the second case the relative change of the density of grain structure was the smallest. The decrease of Δ_{40}, obviously, is mainly connected with increasing the disordered orientation of oblongness of grains, with absence of fragmentation of grain structure onto the additional sub-structures and with decreasing its density. In this connection the largest relative decrease of values Δ_{40} is observed for alloy Cu-2%Al, in which the disorientation of grains appears and the size and number of large grains increase with simultaneous decreasing the minimum size of grains at significant reduction of grain structure density. For alloy Cu-4%Ni the relative decrease of index Δ_{40} after laser treatment is not significant, since the disordered orientation of grains and density of grain structure change insignificantly, the changes of fragmentation of grains onto groups in shape and size is absent.

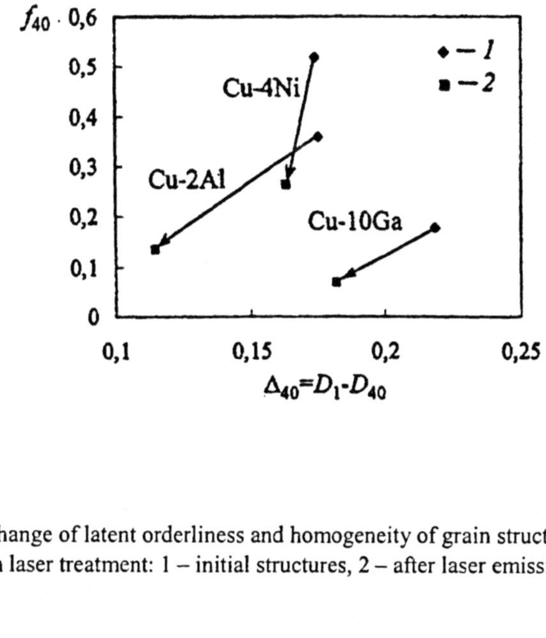

Figure 4. The change of latent orderliness and homogeneity of grain structures of copper alloys resulting from laser treatment: 1 – initial structures, 2 – after laser emission treatment.

The Evaluation of Possible Changes of Mechanical Properties of Material in Zone of LE Treatment

It is obvious that considered LE treatment of material leads to worsening its capacity for work resulting from localization of the tensions and deformations under mechanical load in the treatment zone, where the thickness decreases and quality of surface becomes worth. In this connection for evaluating the capacity for work of material LE treatment it is necessary, in first, to evaluate the mechanical properties of material exactly in the zone of this treatment. For such a evaluation, one can use very near similarity between general character of structure changes caused by LE treatment (the fusion followed by epitaxial mechanism crstallization), with the process of collecting recrystallization. For example, during the collecting recrysallization of molybdenum, considered in work [30], the transition from fine-grain structure appeared during the primary recrystallztion to the structure consisted of large elongated grains with insignificant number of small grains, is also followed with decreasing the homogeneity of structure and growing of its chaotic component; besides, there were also observed the oriented nucleation and growth of grains. Starting from the revealed changes of the traditional and multifractal characteristics of grain structure caused by laser treatment and also taking into consideration the analogy with the processes of secondary recrystallization, one can assume that mechanical properties of melted part of alloy are characterized by lesser strength and larger plasticity in comparison to the initial structure. That leads to more localization of strains and deformation under mechanical load of material in zone of such parts. The most negative structure changes and, respectively, more larger negative influence of parts of these changes upon capacity for work of material arise in alloy Cu-2%Al. The obtained results can be useful for choosing the copper-based alloys (bronze, for instance) as the constructive material in the cases when there is probability of LE treatment.

Basis Correlations Between Changes of Studied Grain Structures and Processes of their Formation Resulting from LE Treatment, Discovered by Using Multifractal Concept

There was revealed the practical identity of relative change of D_q and α_q ($q \geq 1$) for all studied grain structures after treatment under the same technological conditions despite of the visible differences of these structures in both the initial state and after LE treatment. That is the evidence of effectiveness of using multifractal parameterization of structures as the tool of image identification. It is noteworthy that similar correlation between conditions of obtaining the grain structures and traditional parameters (grain size, the density of grain boundaries, etc) is absence.

The use of multifractal concepts as the additional analysis technique for describing the grain structure before and after LE treatment has allowed to discover some main behaviors of their formation, the detection of which without such complex approach would had been very difficult. It has been determined that general behaviors of influencing the alloying elements on the size and shape of grain as well as on the shape of grain boundaries during the formation of grain structure resulting from laser treatment in general are the same as for formation of initial structure. Using only traditional techniques, it is difficult to discovery that behavior, as during the laser treatment of zone the specific "patelous" structures, which strongly differ from the initial structures with shape and sizes of grain, have been formed in the studied alloys.

Furthermore, it has been revealed that structural changes, caused by LE treatment, for all the studied alloys are in general of the same character and are followed with decreasing the homogeneity and orderliness in treatment zone. Such a knowledge about general character of changes of grain structure as the system, which can not be obtained from analysis only traditional characteristics of grain structures, allow determining the analogy between the studied changes and other possible evolutional processes in grain structure. In a given concrete case there was determined the analogy between processes of crystallization after laser through penetration and processes of collective recrystallization. By determining the analogy due to character of change of multifractal characteristics, one can approximately evaluate the direction of change of material mechanical properties resulting from structural changes. That is especially important in the cases when the evaluation of mechanical properties in zones of structural changes is very difficult when using traditional techniques because of small sizes of section or other reasons.

CONCLUSIONS

1. The use of multifractal conceptions allowed significantly amplifying the traditional techniques of describing the grain structures and to reveal some unrevealed grain structures formation behaviors non-observed at traditional approaches under laser treatment.
2. The structural changes, caused by exposure of laser emission upon thin plates of alloys Cu-4%Ni, Cu-10%Ga and Cu-2%Al are of the equal character and are accompanied by the same changes of multifractal characteristics: the decrease of homogeneity and latent orderliness of grain structures in treatment zone. The general behaviors of influence of alloying components upon formation of grain structure under laser treatment were the same as under initial alloy treatment.
3. The methodology of multifractal analysis allowed discovering the similarity of formation of grain microstructure for different conditions of materials treatment – non-equilibrium process of "laser hardening" (petalous grain structure in studied copper-based alloys) and quasi-equilibrium process of collective recrystallization of molybdenum (mainly the elongated grains).
4. They were evaluated the changes of mechanical properties of treated copper alloys and shown that in crystallization zone after laser treatment the strength decreases, and the elasticity increases in comparison to initial state of materials.

The work has been supported by the grant RFFI #98-01-00009a and international grant of European Community INCO COPERNICUS # ERB IC15 CT98 0811.

REFERENCES

1. La Rocca A.V. Laser Application in Manufacturing. Rapport Injorsvelenskapsakad, 1984, no. 258, pp. 31-53.
2. Poale J.M., Foti G., Jakobson D.C. Surface Modification and Alloying by Laser, Ion, and Electron Beams., New York: Plenum Press, 1983, 424 p.
3. Maslyaev S-A., Platov Yu.M., Pimenov V.N. Fizika i khimiya obrab. materialov, 1990, no. 4, pp. 9-13.
4. Maslyaev S.A., Neverov V.I., Pimenov V.N., Platov Yu.M., Betsofen S.Ya-, Sasinovskaya I.P. Fizika i khimiya obrab. materialov, 1992, no. 5, pp. 38-45.
5. Ivanov L.I., Maslyaev S.A, Pimenov V.N., Sasinovskaya I.P., Tsepelev A.B., Nikel X., Linke Dzh., Duve R. Perspektivnye materialy, 1996, Khe 4, pp. 35-46.
6. Maslyaev S.A., Pimenov V.N., Platov Yu.M., Demina E.V., Betsofen S.Ya., Gribkov V.A., Dubrovskiy A.V. Perspektivnye materialy, 1998, no. 3, pp. 39-47.
7. Ivanova V.S., Balankin A.S., Bunin I, Zh., Oksogoev A.A. Sinergetika i fraktaly v materialovedenii, M.: Nauka, 1994, 383 p.
8. Vstovsky G.V. Foundations of Physics, 1997, v. 27, no. 10, pp. 1413-1444.
9. Vstovskiy G.V., Kolmakov A.G., Tsrentev V.F. Izv. Akademii nauk. Ser. Metally, 1993, no. 4, pp. 164-178.
10. Vstovskiy G.V., Bunin I.Zh., Kolmakov A.G. i dr. Doklady RAN, 1995, t. 343, no. 5, pp. 613-615.
11. Vstovskiy G.V., Kolmakov A.G. Fizika i khimiya obrabotki materialov, 1995, no. 6, pp. 69-84.
12. Kolmakov A.G., Geminov V.N,, Vstovsky G.V. et al. Surface & Coating Tcchnol. 1995, v. 72, pp. 43-50.
13. Kolmakov A.G., Bunin I.G., Vstovsky G.V. International Journal of Offshore and Polar Engineering (UOPE), 1997, v. 7, no. l, p. 44-47.

14. Bunin I.Zh., Kolmakov A.G, Vstovskiy G.V., Izv. RAN, Ser. Metally, 1998, no. 1, p. 103-106.
15. Pimenov V.N., Maslyaev S.A., Betsofen S.Ya. Fizika i khimiya obrabotki materialov, 1989, no. 4, s. 24-29.
16. Ivanov L.I., Pimenov V.N., Yanushkevich V.A. Fizika i khimiya obrabotki materialov, 1989, no. 3, s. 41-46.
17. Hentschet H-G.E., Procaccia L.P. Physica D, 1983, v.8,
18. Halsey T.C., Jensen M.H. KadanoffI.P. et al. Phys. Rev. A, 1986, v. 33, no. 2, p. 1141-1151.
19. GionaM., Piccirili P., Cimagalli V.J. Phys. A, 1990, v. 24, no. 1, p. 367-373.
20. Ivanov L.I., Maslyaev S.A., Pimenov V.N,, Yanushkevich V.A. Fizika i himiya obrabotki materialov, 1988, № 2, s. 65-70.
21. Grigorovich V.K. The Metallic Bond and the Structure of Metals. Commack, New York: Nova Science Publishers, 1989, 311 p.
22. Mandelbrot V.V. An introduction to multifractal distribution functions. In: H.E.Stanley, N.Ostrowsky eds. "Random Fluctuations and Pattern Growth: Experiments and Models", Dordrecht, Kluwer Academic, 1988, p. 279-291
23. Jensen M.H., Kadanoff L.P, Procaccia I. Phys. Rev. A, 1987, v. 36, no. 3, p. 1409-1420.
24. Cosenza M.G., Swift J.B. Phys. Rev. A, 1990, v. 41, no. 12, p. 6615-6620.
25. KatzenD., Procaccia I. Phys. Rev. Lett., 1987, v. 58, no. 12, p. 1169-1172.
26. Feigenbaum M.J. J. Stat. Phys., 1987, v. 46, no. 5/6, p. 919-924.
27. Stepfalusy P., Tel T., Csordas A., Kovacs Z. Phys. Rev. A, 1987, v. 36, no. 7, p. 3525-3528.
28. Bohr T., Jensen M.H. Phys. Rev. A, 1987, v. 36, no. 10, R. 4904-49G5.
29. Rangel R., Guerrero L.E. Phys. Rev. A, 1991, v. 43, no. 2, p. 669-680.
30. Kolmakov A.G, Bunin I.Zh., Kozitskiy D.V. Metally, 1999, № 1, s. 80-87.

ABOUT THE AUTHORS

Aleksey Georgievich Kolmakov *– Institute of Metallurgy and Materials Science after A.A. Baikova of RAS. A Candidate of Technical Sciences, research officer. An expert in materials science area, materials mechanical properties, surface effects.*

Grigoriy Valentinovich Vstovskiy *– Institute of Metallurgy and Materials Science after A.A.Baikova of RAS. A Candidate of Physics and Mathematics, research officer. An expert in area of multifractal analysis, destruction mechanics, applied and calculation mathematics.*

Sergey Aleksandrovich Maslyaev *– Institute of Metallurgy and Materials Science after A.A.Baikova of RAS. A Candidate of Physics and Mathematics, research officer. An expert in area of effect of high-power flows upon metallic materials.*

Valeriy Nikolayevich Pimenov*- Institute of Metallurgy and Materials Science after A.A.Baikova of RAS. A Doctor of Physics and Mathematics, leading research officer. An expert in area of radiation solid state physics and space materials science.*

HIGH TEMPERATURE GLASS-CERAMIC REFLECTING COATINGS

M.M. Mikhailov, V.I. Vereshchagin and S.V. Smirnov

The effect of pulse treatment by xenon lamp emission upon structure, reflectance and mechanical strength of magnesia-silicate glass-ceramic reflecting coatings with the different amount of zinc oxygen has been studied in this work. It has been determined that due to zinc oxide the reflection ability increases; the pulse heat treatment causes the formation of glass coatings, increasing the mechanical strength of the coatings.

INTRODUCTION

The reflecting coatings, based on zinc oxide pigment and silicate glass binding material, are widely used in aviation and aerospace engineering. These coatings have low mechanical strength and are not able to be subjected to high-temperature loads. One of the solution of this problem is to replace the silicate glass by a more heat-resistance material thus zinc oxide to be contained in chemical composition of this binding material.

The magnesia-silicate glass-ceramic materials, containing the zinc oxide as the component, are the most promising ones for these purposes. Having a high chemical resistance and heat resistance, such coatings can be used as both the thermo-regulating and the protecting structural units from the exposure of combustion product and intensive infra-red radiation, emitted by jet and turbo-jet plants [1].

The main limitation of reflecting ceramic coatings deposited on the protecting surfaces by means of the smearing and the sputtering is the low mechanical strength and badly reproducible reflecting properties. The elimination of this limitation is possible by means of high-temperature heat treatment that is not always possible for large sizes, irregular shape of surface area and low heat resistance of structure material. The heat treatment of the coatings by intensive IR-radiation of halogen and xenon gas-discharge lamps is the most promising method [2]. Under the treatment, the reversible and irreversible failure processes take place, which can lead to both the positive and the negative consequences. First of all, it is connected with high brittleness and low resistance of ceramic coatings to the thermo-blow.

Therefore the development of reflecting glass-ceramic coatings by means of varying the content and regimes of pulse IR-treatment has scientific and practical interest.

THEORETICAL ASPECTS

Let us consider the process of heating the surface of ceramic material by pulse-periodical irradiation of gas-discharge lamps. The heat blow on the surface can be set by sequence of square pulses of duration of τ and period of Π. It can be mathematically defined by Fourier series:

$$q(t) = \sum_{m=1}^{\infty} \left\{ 4/\Pi \left[q_0 (1-R) \int_0^{\infty} \cos \omega t dt \right] \right\} \qquad (1)$$

Where m=1,2,3...; ω=2mπ/Π; q_0 is the flux density of optical radiation; R is the reflectance. The solution of non-stationary thermal conductivity equation:

$$\frac{dT}{dt} = aDT + q(t) \qquad (2)$$

for r=0, is:

$$\Delta T(0,t) = \sum_{m=1}^{\infty} \left\{ [q(t)/\cos \omega t](1/\lambda)\sqrt{2a\varpi}\sin(\omega t) \right\} \qquad (3)$$

where a and λ are the coefficients of temperature conductivity and thermal conductivity of coating material, respectively.

The graphical solution of equation (1-3), giving the imagination of the heating dynamics of glass-ceramic coating, is presented in Fig.1. If for the time between two neighbor pulses the temperature does not become equal to the coating thickness, then to the beginning of the following pulse the temperature field will be defined by superposition of two components, corresponding to the exposure of radiation source and relaxation heat field. At the end of certain time, corresponding to the crossing of several pulses, the quasi-stationer heat regime starts, in which the temperature field is replicated under exposure of separate pulses.

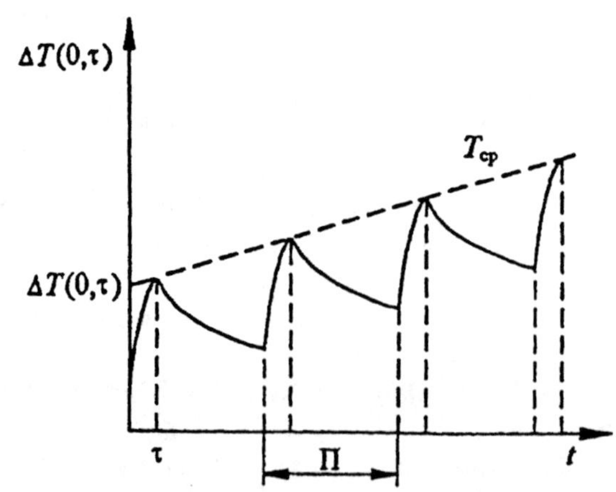

Figure 1. The time dependence of temperature of surface for pulse heating.

The experimental investigations have shown that temperature of the coating surface, treated by single pulse of energy of 10 J/cm^2 and duration of 10^{-3} s, increases for 100-150 degrees. As the number of pulses increases, T_{max} trends to the saturation, the difference between T_{max} and T_{min} decreases. Thus during pulse-periodical exposure of IR-radiation there is the possibility of smooth decreasing of temperature down to a given value.

THE EXPERIMENT

The alum-magnesia-silicate glass-ceramic materials were chosen as the subject of inquiry. These materials are characterized by the stable composition and minimum amount of non-controlled dopants. They are: the forsterite ceramics, steatile ceramics, the forsterite glass-ceramics, steatile glass-ceramics. The chemical composition of materials is given in Table 1.

Table 1: The Chemical Composition of Investigated Material

Type of material	Chemical composition (mass. %)
Talc	MgO - 28; SiO_2 – 54.22; BaO - 15; Al_2O_3 + TiO_2 – 2.78
Forsterite	MgO –47.8; SiO_2 – 43.5; BaO – 5.8; Al_2O_3 – 2.9
Steatile glass-ceramics	MgO - 28; SiO_2 – 58; BaO - 6; Al_2O_3 – 2; ZnO - 6
Forsterite glass-ceramics	MgO - 18; SiO_2 – 36; BaO - 8; Al_2O_3 – 2; ZnO - 36

The investigated materials were synthesized from source oxides by the sintering technology. The powder was thoroughly mixed in ball mill to the size of 3 μm. The obtained powder was mixed with isopropyl alcohol and then was deposited by pulverize onto the metallic tape of the alloys of type of the "cover" and was dried in nitrogen stream for temperature 1000 K, for 20 hours. Simultaneously, from the powder, by means of dry compression the compact specimens of the plates of dimension of 30×24 mm and thickness of 1 mm, which were annealed in silicate furnace at temperature 1500-1600 K during 10 hours, were produced.

The specimens were tested for the mechanical strength (the bending), for the crack resistance (by means of hardness indentation method), as well as was studied by means of reflecting optical IR-spectroscopy, X-ray analysis, prior and after multi-pulse treatment by radiation of xenon gas-discharge lamps. The treatment was carried out under the following conditions: the pulse duration of 10^{-3} s, exposure duration of 10 s, the radiation energy density of 10 J/cm^2 and the succession interval of 5 Hz.

The X-ray analysis has shown (Fig.2) that the main crystalline phase of studied materials is the forsterite and the talc. Aluminum oxide and boron oxide do not practically form the crystalline phases, but are in amorphous state – glass phase. Zinc oxide is predominantly in amorphous phase as well. The observed slight diffraction peaks on the X-ray pictures were interpreted by us as zinc metasilicate. After multi-pulse treatment in the talc, forsterite and steatile glass-ceramics, the portion of amorphous phase increases that are characterized by weakening of diffraction peaks. But in the coating based on the forsterite glass-ceramics the secondary forsterite crystallization is observed that is confirmed by increase of diffraction peaks, corresponding to forsterite, on the X-ray picture.

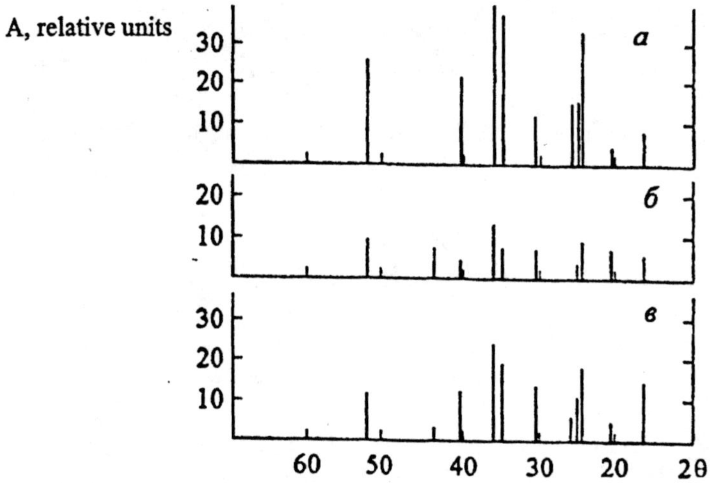

Figure 2. The bar-diagrams of diffraction peaks of the coatings: *a* – forsterite; *b* – forsterite glass-ceramics; *c* – forsterite glass-ceramics after multi-pulse treatment.

The analysis of reflectance IR-spectra of studied materials (Fig.3) before and after pulse treatment shows that on the surface of ceramic and glass-ceramic coatings, treated by pulse radiation, the glass phase of 10 μm thickness is formed. In the talc and forsterite IR-spectra there is observed the increase of absorption intensity of band due to decrease of surface roughness. In reflectance IR-spectra for forsterite and steatile glass-ceramics there is observed the decrease of intensity of absorption band of 405-410 cm^{-1}, corresponding to oscillations of groups $[Zn^{2+}O_6]$ that substantiates the assumption of partial evaporation of zinc from surface of glass-ceramic coatings. In reflectance IR-spectra for forsterite glass-ceramics after pulse treatment there is observed the significant increase of intensity of absorption band at 432 cm^{-1}, corresponding the valency oscillations of groups $[Mg^{2+}O_6]$ in structure of forsterite that confirms the above conclusion about secondary crystallization of the forsterite.

The investigations of diffusion reflection spectra in visible region have shown that the largest reflectance is inhered to forsterite glass-ceramics, containing 36% of ZnO. It precedes 80% in region of 450-500 nm, being 70% of whole visible region.

The coefficient of diffusion reflection R_{dif} of coatings in range of wavelength from 400 up to 750 nm after multi-pulse treatment (Fig.4), as rule, decreases, that indicates the increase of mirror component of reflection. For the coatings, based on forsterite glass-ceramics, the change of R_{dif} is irregular in spectrum; in the region near 470 nm there is observed the peak, which is probably connected with firmed defects. As the given composition is distinguished with large amount of zinc (36%), the appearance of absorption band at 470 nm may be conceivably connected with defects in ZnO. The vacancies of zinc and the near pairs of Frenkel [5] may be the such defects that confirms the early made conclusions about nature of decrease of intensity of band 405-410 cm^{-1} in IR-spectra (Fig.3). It is possible that the increase of concentration of vacancies of zinc is caused not so much by IR-radiation as by the exposure of short-wave radiation of visible and UV-regions present in spectra of pulse xenon lamps.

The amorphousing and fusing of the surface of ceramic and glass-ceramic coatings causes the healing of surface structure defects that influences upon increasing the bend breaking point of materials and the coefficient K_{1c} of failure viscosity, measured by means of indentation of diamond indenter (Vikkers' pyramids). The results of measurements are presented in Table 2. The increase of the strength is explained by both the decrease of structure defects on the surface and the appearance of sconstringent stresses in thin layer of recrystallized glass phase, which cause the essential influence on distribution of internal stress in plate.

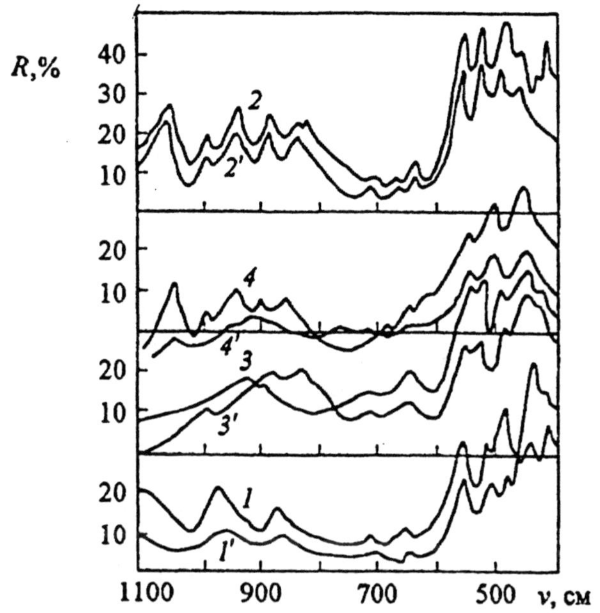

Figure 3. IR-spectra of surfaces. 1, 1' – forsterite glass-ceramics; 2, 2' - Steatile glass-ceramics 3, 3' – forsterite; 4, 4' – talc. The index ' relates to the surfaces subjected to multi-pulse treatment.

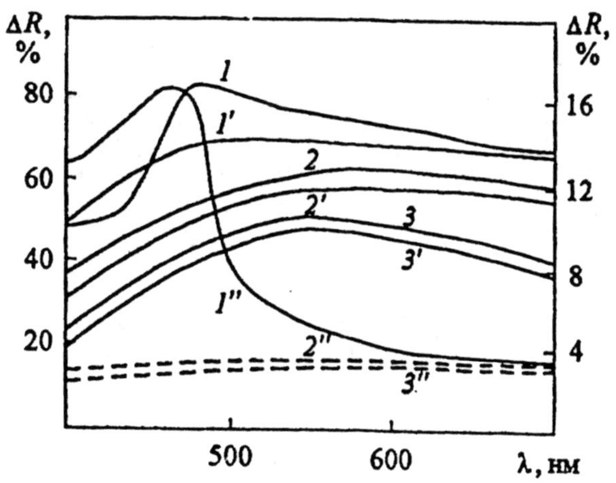

Figure 4. The spectra of diffusion reflection of surfaces in visible region. The indications on the dependencies are the same as in Fig. 3. The index " relates to the differential spectra of diffusion reflection.

Table 2: The Change of Strength of Ceramic Specimens at Multi-Pulse Treatment

Material	% of glass phase	Strength, Mp	Increasing the strength, Mp	K_{1s}, Mp·m$^{1/2}$	ΔK_{Is}, Mp·m$^{1/2}$
Talc	25	140	20	1.4	0.5
Forsterite	15	140	20	1.46	0.64
Steatile glass-ceramics	25	160	20	1.6	0.6
Forsterite glass-ceramics	35	155	28	1.55	0.35

Thus the resilience in layer of glass phase can be defined as follows:

$$\varepsilon_{zz} = \frac{1+\mu}{3(1-\mu)}\left[\Omega(z) + 3\alpha_T \Delta T\right] \qquad (4)$$

where Ω(z) is the stress function, caused by recrystallization processes in layer of glass phase.

In general, the elastic stresses in plate σ(*t*) depend on both the time pulse treatment and the passed time. They consist of the thermoelastic deformations σ_{therm} and the stresses σ_{recr} caused by recrystallization. The estimation of thermoelastic deformations was performed above, and the calculation of stresses, appeared at recrystallization, is performed with using the Mazurin's method [3]:

$$\sigma(t) = \sigma_{Term} + \sigma_{rekr} = -E\frac{\Delta l}{l}\exp\left(-\frac{\sqrt{\xi_k}}{\tau_{\sigma r}}\right) \qquad (5)$$

where $\Delta l/l$ is relative reduction of glass phase layer for recrystallization; ξ_k is some constant, being of time dimension; $\tau_{\sigma r}=\varphi/K_\sigma$; η is viscosity of glass phase, K_σ is the coefficient of proportionality. From the equation (5) it follows that in near-surface layer, the stresses have constringent character. It is connected with fusing the surface and its following solidification, as well as with difference of coefficients of thermal expansion of glass phase and initial material. The stresses caused by only these conditions can be defined from expression:

$$\sigma_{c\phi} = \frac{E_{c\varphi}\Delta T(\alpha_{Tsf} - \alpha_{TKer})}{(1-\mu_{C\phi})(1 + D/2h)} \qquad (6)$$

where ΔT is the glass phase vitrifying temperature; index "sf" relates to thermomechanical characteristics of glass phase.

The glass phase layer essentially decreases the concentration of mechanical defects on the surface, as well as effects analogously to that observed at glass glazing of ceramic surface or its strengthening with the help of ion exchange [4].

In general case, the processes of strengthening of ceramics with help of pulse treatment can be reduced to the following mechanisms.

1. The relaxation in near-surface layer of elastic stresses, remained after mechanical and heat treatment.
2. The formation of diffusive flow of defects from the surface into the depth of material due to temperature gradient (Sore's process) and completion of crystalline structure ingrains.
3. The surface fusing of glass phase, its interaction with crystalline phase and completion by the glass phase of surface pores and cracks.
4. Appearance of elastic stresses in near-surface layer of elastic stresses, caused by fusion processes and solidification of glass phase.

As followed from experimental results, the role of glass phase during the strengthening is the determinative one.

CONCLUSION

The performed experimental investigations and the carried out calculations have shown that glass-ceramic materials, based on magnesia-silicate, with zinc oxide dopants can be used as the high-temperature reflecting surfaces for IR- and visible spectrum region.

The multi-pulse IR-treatment of glass-ceramic surfaces causes the formation of thin glass layer that leads to increasing their mechanical strength and crack resistance. For the coatings with zinc oxide dopants, the mechanical strength is significantly larger.

The reflectance of coatings after multi-pulse treatment slightly decreases that can be connected with increasing the mirror component of reflectance coefficient and formation of proper point defects.

REFERENCES

1. Mikhailov M.M., Dvoretskii M.I. Journal of Advanced Materials, 1995, no. 2(l), p. 41-49.
2. Romanov B.P., Smirnov S.V., Aleutdinov A.D. Modifitsirovanie svoystv glazuri i steklofazy keramiki elektrofizicheskimi istochnikami nagreva. Tez. dokl. 2 sъezda keramicheskogo obschestva. Moskva, 1991, s. 46-47.
3. Mazurin O.V. Otzhig spaev stekla s metallom. Energiya. Leningrad, 1980, 140 s.
4. Loges F., Christ A., Seifert W. Eigenspannungen in Glasartigen Systemen mit Kontinuirlicher Oberflachennaher Quellverteilung. FMS-ser. Anst. Mech. Akad. Wiss. DDR. 1987, D. 29, 49 s.
5. Mihaylov M.M., Sharafutdinova V.V. Izv. VU3ov. Fizika, 1997, № 9, s. 106-112.

ABOUT THE AUTHORS

Mikhail Mikhailovich Mikhailov– *The Tomsk Polytechnic University. A Professor. A Doctor in Physics and Mathematics, Head of Chair. An expert in area of radiation physics, radiation materials science and materials for space vehicles.*

Vladimir Ivanovich Vereshchagin – *The Tomsk Polytechnic University. A Doctor of Technical sciences, Head of Chair. An expert in area of ceramics and glass technologies.*

Serafim Vsevolodovich Smirnov – *The State Plant Science Research Institute "Semiconductor devices". A Doctor of Physics and Mathematics, Head of Chair. An expert in area of laser technologies.*

Peculiarities of Obtaining and Studying the Platinum Emitting Compounds

G.G.Bondarenko, S.M. Zhdanov and A.P. Korzhavyi

We present the results of obtaining and studying the emission properties of compounds based on the platinum with oxides BaO, MgO, SrO, Al_2O_3, BeO, CaO, Y_2O_3 and SiO_2. The optimal regime for activating such compounds is: T=1270 K, t=1 h. That provides the maximum coefficient of secondary electron emission from platinum compounds within 3.1 and 4.5 and first critical potential within 31.5 and 18.5 eV. It has been shown that investigated platinum compounds do not essentially influence on the treatment and field emission of tantalum cold cathode in common cathode unit. The processes during activation and exploration of cold cathodes in experimental auto-start vacuum devices have complex and constantly changing character, depending on condition of surface of cold cathode.

Introduction

The development of new emitting materials, first of all the secondary-emission ones, having the high and stable coefficient of secondary electron emission σ (CSEE) and then first critical potential E^I_p, under electron bombardment, is connected with development of whole class of vacuum microwave (MW) amplifiers and generators. Such a secondary-emission cathodes must be explored and must maintain its parameters in range of energy of electron, bombarding the cathode from 50 up to 1000 eV and in wide temperature interval. The prospective materials, which are able to meet these requirements, are the compound materials of type of metal-oxides [1-3], which are explored together with cold cathodes. There were industrially produced the secondary-emission cathodes, obtained by powder metallurgy methods, when the powders of tungsten, rhenium, palladium and nickel are used as the substrate, and the oxides of alkaline-earth metals are used as the active dopants [4]. They have the essential advantages in exploration under conditions of power electron bombardment, which are connected with that every grain of active component works as individual emitter from the dielectric (oxide or blend of oxides), supplying the electron from surrounding metallic matrix [1]. As the grains of active component of compounds are large enough, their destruction under equal current densities passes slower in comparison to analogies, produced from activated metallic alloys and the accumulation of metallic products of dissociations of active component decreases with increasing the temperature [5].

The high secondary-emission properties, inhered to compound materials, based on the palladium, with using complex compounds of oxides of alkali-earth metals [5], gives the argument to affirm that the direction of production composition production with using the noble metals is one of the most promising and the possibilities are far from exhausting.

The aim of present work is to obtain and study the secondary emission and other properties of platinum-based materials. The complex blends and oxides compounds BaO, MgO, SrO, Al_2O_3, BeO, Y_2O_3 and SiO_2 were used as active dopants.

Obtaining of Specimens and the Investigation Techniques

The experimental specimens were produced by powder metallurgy method according to technological scheme: making the active mixture – making the composition working mixture – pressing the preform – sintering the specimens – mechanical treatment.

There was used the platinum PPl-II containing 99.8% of basis material. Platinum specimens contain iron of amount of 0.05 mass. %; 0.006 mass% of silicon; 0.006 mass.% of lead. By means of additional treatment the amount of dopants in platinum was minimized. The size of grains (particles) size is 30-70 μm.

For obtaining the active components of compound, there was used beryllium oxide, yttrium oxide, magnum oxide and silicon oxide. The rest oxides were obtained from carbonates. The size of particles of oxides is 3-5 μm and lesser. For the purpose of maximum purification from dopants, they all were additionally treated. The typical ratio of oxides: BaO:MgO:BeO=3:5:6; MgO:BeO:Al_2O_3=1:1:2; BaO:CaO:Al_2O_3=2.5:0.5:1; BaO:CaO:BeO:Al_2O_3=2.5:0.5:0.5:1 and other.

The mixture was thoroughly mixed in vibratory mill where the metallic and oxide particles were additionally grinded. After one hour mixing, the mixture was replaced from mill and placed into desiccator. Experimental composition specimens were obtained by means of cool pressing in steel die molds for specific pressure of 5 t/cm^2 followed by sintering at temperatures 1070-1370K, depending on composition of oxide phase. By means of mechanical treatment the compound specimens possessed the required working surface.

The indispensable condition of obtaining the high emission properties and resistance to electron bombardment is the optimization of heat treatment of composition materials, i.e. the emitting compound must be activated. The optimal regime was determined from dependence of CSEE and temperature and activation duration of experimental emitting specimens.

The determination of dependence of CSEE vs. energy of primary electrons E_p, value E^1_{p} and the test for durability of platinum compositions was performed in vacuum tubes on the plant widely used for these purposes [6,7] using special equipment [8].

The study of tantalum blade cold cathode of thickness of 4 mm was performed with surrounding the platinum secondary-emission compounds. The technique of these investigations is presented in works [9,10]. The peculiar features of these investigations is that in experimental electro vacuum device the tantalum foil cathode is the ring of d_{int}=4.6 mm, d_{ext}=6.2 mm, it was mounted on hollow molybdenum core between two platinum secondary-emission specimens of 5.9 mm in external diameter, i.e. the height of external edge of tantalum field-emission-cathode is 0.15 mm.

For heating such emitting cathode unit to the necessary temperature (not lower than 1200 K) the spiral tungsten heater was embedded into the hollow of molybdenum core. Temperature was controlled by means of chrome-nickel thermocouple. The assembled unit was coaxially embedded into ring anode of gap of anode-auto-cathode of 1.2 mm, according to the requirements, presented in work [10]. The gap of width of 0.4 mm was broke through, placed along generatrix that allowed spinning it around coaxial axis with the help of external magnet. The last detail of experimental device is the collector, surrounding the anode outside.

Collector and anode of experimental device are connected with common conduct through micrometer I_1 and I_2 and they have equal potential. The potential difference between emitting cathode unit and common conductor was set by means of regulating anode source U_a from possible voltage failures in vacuum space, the anode source was switched to experimental device via ballast resistance R_b. The field-emission electrons from the part of the edge of length of 0.4 mm, placed against the gap in cathode, pass through, fall onto collector and cause the current I_1 in anode chain. From the rest part of tantalum edge the field-emission electrons produce the current in anode chain I_2.

Thus produced experimental device allows determining both the distribution of field-emission current vs. length of field-emission tantalum edge (at turning the anode around axis) and the total current $I=I_1+I_2$. It should be noted that potential difference was chosen experimentally from condition of obtaining the reproducible dependencies $I=f(U_a)$, where I is the total current when the collector is shorten on anode. Besides, for avoiding the damage of cold cathode in the investigated field-emission unit, the maximal value U_a in static regime was limited by 6 kV.

The coefficient of activation of cold cathode in contact with platinum secondary-emission composition K was determined as the emission current of cold cathode I_a after heating-through with respect to primary emission current I_i.

In order to determine K at heating the cold cathodes contacting to the secondary-emission specimens in single unit, in every experiment there was determined the primary current of field-emission after the degassing of unit at 1170 K prior to achieving the pressure of $8\cdot10^{-7}$ mm of mercury ($1.06\cdot10^{-4}$ p). In device the rate of increasing the temperature of degassing was limited by demands required to vacuum devices: the pressure must not exceed $1\cdot10^{-6}$ mm of mercury ($1.33\cdot10^{-4}$ p). The same requirement remains when temperature increases during activation of composition cathode unit with cold cathode.

INVESTIGATIONS OF MAIN EMISSION PARAMETERS OF PLATINUM COMPOUNDS AND A DISCUSSION OF THE RESULTS

In Fig.1 the dependence of maximuml CSEE σ_{max} and first critical potential E_p^I on. temperature T_a and time activation t_a of some listed above platinum compositions is presented. It is seen that during activation both the CSEE and the E_p^I changes. In the range of temperatures of 720-770K, at heating the compounds, the intensive release of H_2O, CO_2 and CO is observed. The release rate of these gases sharply increases, σ_{max} increases and E_p^I decreases with temperature. Increasing the activation temperature above 1170 K, the secondary-emission properties for platinum compositions, presented in Fig.1, are worsened (the change of secondary-emission properties was performed at T_{maes}=870 K). This is connected with that, when increasing the evaporation rate of oxide phase, the composition of surface changes, the emission centers increase and surface is recharged in the result of influence of these centers.

The results of investigation of secondary-emission properties of some emitting platinum compositions are presented in Table 1.

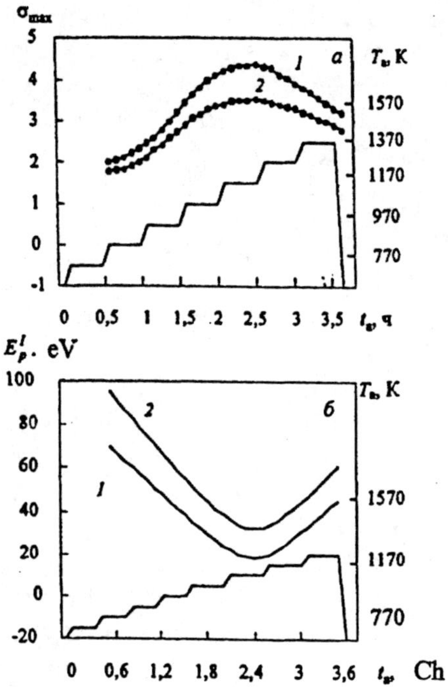

Figure 1. The temperature-time dependence $\sigma_{max}=f(T_a,t_a)$ (*a*) and $E_p^I=f(T_a,t_a)$ (*b*) at T_{meas}=870K: 1 – Pt+4.6% (3BaO·5MgO·6BeO), 2 – Pt+5% (2.5SrO·0.5CaO·Al_2O_3).

It should be noted that, first of all, the temperature-time curves of activation of studied compositions have the mildly slopping maximum in temperature range of 1070-1370 K and dependence $\sigma=f(E_p)$ has the mildly slopping maximum in range of primary electron energy of 300-1400 eV. One should suppose that optimal activation temperature for these specimens is T_a=1170 K, as that ensures high-emission parameters (see Table 1). It was noted that the heating of composition up to temperatures 1280-1370 causes decreasing the CSEE (not more than for 5-6%). Most probably, when using these compositions as secondary-emission cathodes, it is necessary to explore them below 1370 K, ensuring σ_{max}=3.15-4.51 and E_p^I=18.5-31.5 eV. The secondary-emission parameters of platinum compositions, besides, have the evidently expressed temperature dependence (the temperature coefficient is positive for CSEE and is negative for E_p^I)

The resistance to electron bombardment of investigated compositions is studied for the density of j=40 mA/cm^2 in vacuum tubes in static regime of exploration of electron gun [8] for continuous taking the secondary electron current out by means of analysis of dependence $\sigma_{max}=f(t)$. The results of these investigations are presented in Fig.2. It is seen that after 2500 hours of electron bombardment of platinum compositions at working temperature of 970 K, the maximum coefficient of secondary emission decreased down to 2.6-3.1 that for the power of electron bombardment of 40 W/cm^2 was not successfully obtained for other emitting materials [8]. The decrease of CSEE in the process of long-duration electron bombardment of platinum compositions can be explained by constant change of chemical composition because of different rates of dissociation of oxides. After thousands hours of electron bombardment, the equilibrium between process of inflow of active composition and its damage under electron bombardment is established, causing the stabilization of secondary-emission properties of investigated compositions. For the duration of 1000 hours, the value σ_{max}, forecasted due to modern techniques, is equal to 2.3-2.5 that meets the up-to-date requirements.

The determination of optimal conditions of activation of tantalum cold cathodes contacting the platinum composition in the only unit was carried out for different temperatures in range of 870-1170 K for every 100 degrees. The measurement of primary current of field emission and current after every activation cycle was carried out for the same temperature of the only cathode unit and the residual pressure of $2\text{-}5\cdot10^{-7}$ mm of mercury ($2.66\cdot10^{-5}$ p) in experimental tube. Usually the time of activation until complete stabilization of total current of field-emission in investigated temperature interval did not exceed one hour.

Table 1: Secondary-Emission Properties of Compounds of Platinum Based Compounds (T_{mes}=870 K)

Compound	Activation regime		CSEE at E_p^I, eV		σ_{max}	E_p^I, eV
	T_a, K	t_a, h	300	400		
1. Pt + 4.6% (3 BaO*5 MgO*6BeO)	1170	1.0	3.87	4.13	4.51	18.5
2. Pt + 4.6% (3 BaO*5 MgO*6BeO)	1070	1.0	3.40	3.75	3.99	25.0
3. Pt + 5% (2.5 SrO*0.5 CaO*Al_2O_3)	1170	0.5	2.97	3.08	3.15	31.5
4. Pt + 5% (2.5 SrO*0.5 CaO*Al_2O_3)	1170	1.0	3.13	3.25	3.38	24.0
5. Pt + 5% (2.5 SrO*0.5 CaO*Al_2O_3)	1070	1.0	3.12	3.23	3.24	26.8
6. Pt + 5% (2.5 SrO*0.5 CaO*Al_2O_3*0.5 BeO)	1170	1.0	3.03	3.15	3.15	29.5

The results of investigations on behavior of field-emission current (in regime of total current) from tantalum cathode in contact with different platinum compounds at optimal temperature are presented in Table 2. Here I_a^x is the field-emission current after the heating through at 1270 K.

As seen in Table 2, for the tantalum cold cathodes with all investigated materials, contacting them, the activation coefficient K is from 5.1 up to 21 after heating in temperature range of 970-1070 K. It is seen that after activation, there is not viewed any behaviors of properties dependencies on the material contacted with cold cathode. Practically, all the materials contacting with cold cathode do not essentially effect on I_n, I_a, T, K, which practically do not differ from the values obtained at investigating the tantalum cold cathode, contacting the molybdenum.

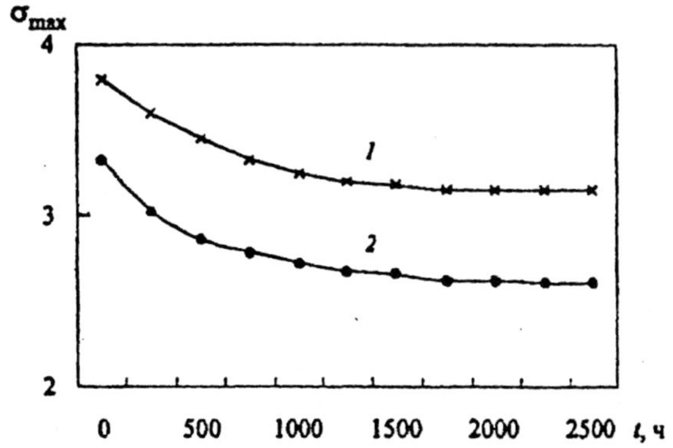

Figure 2. The dependence $\sigma_{max}=f(t)$ at j_p=40 mA/cm^2, T_{wk}=970 K, E_p=1000 eV: 1 – Pt+4.6% (3BaO·5MgO·6BeO), 2 – Pt+5% (2.5SrO·0.5CaO·Al_2O_3).

The heating of units at 1270 K causes the essential decrease of field-emission current in comparison to its maximum value. The observed decrease of current of field emission is typical for all the units, including the molybdenum units as well. Obviously, in temperature range 970-1070 K the process of activation is connected with degassing and purification of tantalum cold cathode from different technological contamination, loosening and bursting of basis oxide Ta_2O_5, covering tantalum in the form of thin film at temperatures near 770 K [10]. After activation at temperatures, including 1270 K, the cold cathodes, in all the cases come off without damage from contacting materials that underlines the fact that cold cathode does not interact with them. The decrease of field-emission ability of cold cathode after heating at 1270 K can be connected with thermal coarsening of edges of cold cathode and is caused by that the free surface energy on the peaks of edges is larger than at basis. As the result of that, the atom flow from the top to the bottom appears. This flow is directed to leveling the free energy that causes "washing out" of the tops. The pointed atom movement becomes essential at critical temperature T_{cr} 0.3 T_{ml} [11], where T_{ml} is the tantalum melting temperature. The values of beginning and following (after activation) current of field emission vs. length of cold cathode edge are presented in Fig. 3. From the experiment it has been determined that the primary currents of field emission can appear on one or two parts of cold cathode edge. In the process of thermal activation, a new emitting centers appear, the existing ones increase, remain the unchanged or decay. Prior to activation, the field-emission currents from emitting centers do not exceed 0.005-0.05 μA, the total current from cold cathode is practically equal to the sum of currents from emitting centers. After activation the sum of currents from emitting centers is equal to 0.3-0.5 of total field-emission current that is caused by increasing the emission ability of all the edge of cold emitter due to length up to 0.1-0.3 μm. The results, obtained from investigation of the units, where the contacting materials were the molybdenum and the platinum compounds Pt-$2Y_2O_3$·3BaO·MgO, in respect with field emission, were practically identical.

The Fig. 4 demonstrates the results of investigations of cold cathode under condition of long current take-off for anode voltage in the diode of above-described construction of 6 kV. From presented diagrams it is seen that currents from emitting centers, created after activation, decrease for 3-4 time with simultaneous appearance of a new, but a less intensive centers. The total current of field emission decreases for 1.5 time that is in agreement with data obtained early in ordinary diode. The typical is that the appearance and decay of emitting centers, at both the activation and the long duration experiment any visible behaviors do not develop. The obtained results indicate that activation of tantalum cold cathodes contacting the secondary emission compounds, in temperature range up to 1270 K, do not effect on the processes occurring on other parts of experimental unit. By comparing the photoelectric work function of electrons from pure tantalum (4.12 eV) and for oxide Ta_2O_5 (4.65 eV), one can assume that emitting centers of cold cathode correspond to those parts where the oxide film is destroyed and the currents of emitting are defined by radii of the rounding of edges of these parts. The residual part of cold cathode edge is covered with tantalum oxide films.

Table 2 Coefficient of Cold Cathode Activation with Secondary-Emission Platinum Compounds and Molybdenum

The material contacting tantalum cold cathode	I_n, µA	I_a,µA	K	T_a, K	I_a^x, µA
1. Pt-$2Y_2O_3$*3BaO*MgO	26	150	5.8	970	60
2. Pt-$2Y_2O_3$*3BaO*MgO	10	210	21	1070	950
3. Pt-$2Y_2O_3$*3BaO*MgO	8	90	11.2	970-1070	50
4. Pt-3BaO*0.6CaO*Al_2O_3*$0.5SiO_2$	13	120	9.2	1070	55
5. Pt-3BaO*0.6CaO*Al_2O_3*$0.5SiO_2$	80	440	5.5	1070	210
6. Pt-3BaO*5MgO*6BeO	16	145	9.0	970	75
7. Pt-3BaO*5MgO*6BeO	12	130	10.8	970-1070	75
8. Mo	18	200	11.1	1070	90
9. Mo	8	130	16.2	970	50
10. Mo	120	610	5.1	970-1070	310

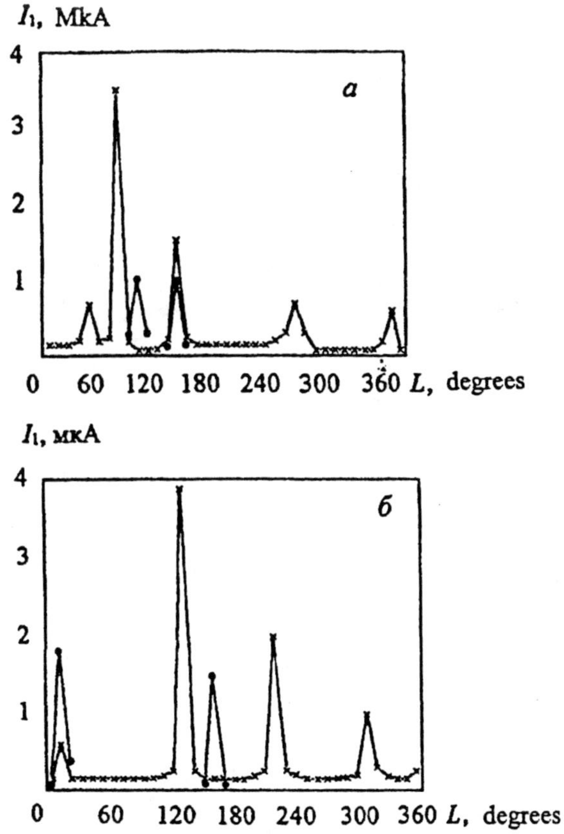

Figure 3. The distribution of the field emission current vs. edge of ring tantalum cold cathode contacting molybdenum (*a*) and compound Pt-$2Y_2O_3$·3BO-Mg (*b*): – prior activation, × - after activation.

The activation of tantalum cold cathode in temperature range 970-1070 K causes its purification from evaporating dopants, loosening and bursting of film Ta_2O_5. Thus the new emitting centers appear on these parts of cold cathode, the rest part of edge of cold cathode, providing the currents of 0.1-0.3 μm vs. the length, remains covered with loosened oxide film. Such a assumption is based on that, on one hand, the Ta_2O_5 is practically insoluble in tantalum and its alloys for investigated temperatures of activation [10], on other hand, in this temperature interval the oxygen dissociation pressure is $1{\cdot}10^{-26}$ mm of mercury ($1.33{\cdot}10^{-24}$ p) [12]. During long duration of current take-off the currents of field-emission from edges decrease gradually due to their bluntness at local overheating, on the parts of cold cathode coated with film Ta_2O_5 at current of 0.1-0.3 μA along edge, depending on degree of loosening of films, the local points with high density of current take-off appear. It causes the damage of oxide films and formation of new emitting centers. The treatments of experimental units do not noticeably effect parameters of platinum compounds: there was not found the traces of evaporation, interaction and decrease of secondary-emission properties in analysis of investigated specimens.

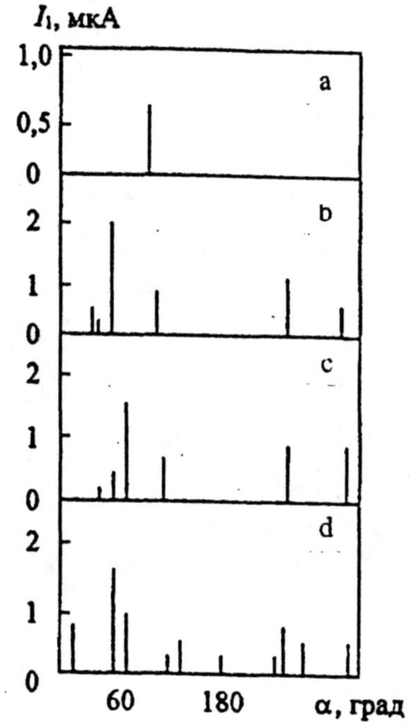

Figure 4. The chart of distribution of emission centers over tantalum cold cathode: *a* – after degassing, *b* – after activation, *c* – after 1 hour of current shunt, *d* – after 10 hours of current shunt.

Hence, for up-to-date vacuum ultra frequency-devices with autostart, the cathode units with platinum compound can be successfully used. Thus the tantalum foil cold cathode can be placed both in the contact with secondary-emission platinum compound and with molybdenum core.

CONCLUSIONS

1. The regimes of obtaining the optimal secondary-emission characteristics (coefficient of secondary electron emission, first critical potential, temperature and activation time) of some

platinum-based compounds with active dopants of alkaline oxides and alkali-earth metals oxides have been determined.
2. At studying the long-time electron bombardment on platinum compounds it has been shown that their secondary-emission properties maintain during $2.5 \cdot 10^3$ hours at least; the forecasted evaluation of their durability is 10^4 hours that is very high index and satisfy up-to-date requirements.
3. The emission properties of the only cathode unit, consisting of tantalum cold cathode and secondary emission platinum compound contacting to it have been investigated. It has been concluded that these cathode units can be recommended for usage in vacuum ultra-high frequency auto-start devices.

The work was financially supported by RFFI and Ministry of Education.

REFERENCES

1. Lepeshinskaya V.N. O nekotorykh tipakh effektivnykh emitterov i ikh prakticheskom primenenii. Tr. LPN. Ser. Fizicheskaya elektronika, 1973, vyp. 328, s. 3-9.
2. Bondarenko G.G., Korzhavyy A.P., Kristya V.I., Rozhkov A.M. Metally, 1993, № 3, s. 101-105.
3. Belov I.K., Bondarenko G.G., Zhdanov S.M., Korzhavyy A.P. Trudy VIII Mezhnatsionalnogo soveschaniya "Radiatsionnaya fizika tverdogo tela". M.: NIIPMT, 1998, s. 377-381.
4. Artsikhovich V.F., Busol F.I., Zvonstskiy V.I. i dr. Elektronnaya promyshlennost, 1984, vyp. 5, s. 50-51.
5. Korzhavyy A.P., Reyaega K.P. Materialy dlya katodov s nizkimi znacheniyami pervogo kriticheskogo potentsiala: Obzory po elektronnoy tekhnike, ser. 6, Materialy. 1987, M.: TsNII "Elektronika", vyp. 2 (1269), 39 s.
6. Bronshteyn I.M., Frayman B.S. Vtorichnaya elektronnaya emissiya. Nauka, 1969, 407 s.
7. Shulman A.R., Fridrikhov S.A. Vtorichno-emissionnye-metody issledovaniya tverdogo tela. M.: Nauka, 1977; 551 s.
8. Korzhavyy A.P. Vtorichno-emissionnye materialy. Metody issledovaniya i prognozirovaniya ikh svoystv: obzory po elektronnoy tekhnike, ser. 6, Materialy, 1984, M.: TsNII "Elektronika", vyp. 11 (1059), 44 s.
9. Savitskiy E.M., Burov I.V., Pirogova S.V., Litaak L.N. Elektricheskie i emissionnye svoystva splavov. M.: Nauka, 1978, 269 s.
10. Niobiy, tantal i ikh oksidy. Pod redaktsiey E.M. Savitskogo. M.; Metallurgiya, 1966, s. 211-237.
11. Sokolskaya I.L. Primenenie avtoemissionnogo. mikroskopa dlya izucheniya poverkhnostnoy diffuzii i samodiffuzii. Poverkhnostnaya diffuziya i rastekanie. Pod redaktsiey Ya.E. Geguzina. M.: Nauka, 1969, 108 s.
12. Fiziko-khimicheskie svoystva okislov. Spravochnik. Pod redaktsiey GV.Samsonova. M.: Metallurgiya, 1978, 260 s.

ABOUT THE AUTHORS

***Gennadiy Germanovich Bondarenko** – the Moscow State Institute of Electronics and Mathematics (Technical University), Professor, Doctor of Physics and Mathematics. An expert in area of solid-state physics.*

Sergey Mikhaylovich Zhdanov *– the Kaluga Branch of MSTU after N.E. Bauman, Candidate of Technical Sciences, an Associate Professor. An expert in area of development of materials for electronics.*

Aleksey Panteleevich Korzhavyi *– the Metallurgical plant "Ametist" (Kaluga), Doctor of Technical Sciences, a Professor. An expert in area of development of materials for electronics.*

Morphology and Physical Properties of Triglycinsulphate and Lithium Niobate Ferroelectric Single Crystals

V.A. Meleshina and N.V. Belugin

The correlation between physical properties and morphology (domain structure, internal bias fields E_{bias} and dislocations, connected with point defects, variation of amount of intrinsic and doping elements) of ferroelectric crystals has been studied. As an example, for water-soluble triglycisulphate ($(NH_3CH_2COOH)_3SO_4$, T_c=49C) and lithium niobate $LinbO_3$:Fe^{3+}, T_c=1170-1210C and T_{mlt}=1260C), it has been shown that domain structure of ferroelectrics, the fine structure and the bias fields are the parameters which allow controlling the growth variations. Based on the investigation by local X–ray micro analysis (LXMA) of the variations of amount of intrinsic (Nb) and doping (Fe) elements, the localization of charge domain walls of kind P_s^+/P_s^+ and P_s^-/P_s^- $LiNbO_3$:Fe^{3+} in crystals has been explained.

Introduction

The ferroelectric crystals, having a high dielectric constant, a large piezo-module and pyroelectric constant and special electro-optical properties, are widely used in radio- engineering, electro-acoustics, quantum electronics and measuring engineering. The ferroelectrics are used for producing the small-size capacitors, the generators of ultra-sonic waves and the waves of quadratic component (for example, $LiNbO_3$), the high-sensitive pyro-receiver of thermal radiation (for example, triglycisulphate (TGS)).

The anomalies of physical properties of crystals in general, and ferroelectics in particular, as rule, are of genetic origin [1], appearing during crystallization. The reconstruction of history of crystal formation and the determination of reason of appearance of anomalies, the study of real structure and composition, followed with increasing the resolution, is required.

For the ideal ferroelectic, the domains are polarized in arbitrary directions, provided that electrostatic intensity is higher than coercive field. For practical usage, it is important to “fix” rigidly the single domain polarized state that must maintain high polarization homogeneity.

The domain walls of ideal ferroelectric must not deflect significantly from polar axis, moreover to be perpendicular to it.

As an example, for triglycinsulphate crystals and lithium niobate ($LiNbO_3$), in a given work, we study not only the reasons leading to disruption of two mentioned properties of ideal ferroelectrics, but also (as an example, for crystals TGS) there is considered the influence of these disruptions upon classic parameters of phase transformation ferro– paraelectric as well as the processes of repolarization (at visualization of positions of domains walls by means of preferential etch method).

The unipolarity is the state of crystal with the preferred and stabilized orientation of domains in one of the direction of spontaneous orientations. The unipolarity (which in literature is named as the natural, in contrast to the induced one under effect of different external physical influences) is one of the anomalies appearing in crystals of water-soluble ferroelectrics while their growing. The reason of its appearance is supposed to be the change of pH of solution and entering of uncontrolled "background" dopants (Si, Ca, Mg).

For using the ferroelectrics in some devices (in memory elements, measuring engineering, as the control elements in automatic stabilization systems) the natural unipolarity is serious barrier. For other applications based on using piezo- and pyroeffect of ferroelectrics, the induced homogenous unipolarity is necessary.* The problem is that obtaining of the induced (post-growing) unipolarity is the overcoming of genetic memory of crystal, causing the growing unipolarity, natural unipolarity, since the latter in the crystal bulk is heterogeneous [2].

The practical usage of ferroelectrics as well as the scientific investigation connected with testing the phenomenological models and behaviors have required the thorough study of both the macroscopic display of unipolarity (behavior of physical properties) and the microscopic investigations of homogeneity of polarization itself. Because of the difficulty of separating the contribution of every of them, the preliminary investigations characterizing the peculiarities of their domain structure, defect structure and homogeneity of composition are of great significance. In general, these characteristics in literature are named as morphologic sub-structures, obviously, in the connection to that any part inside crystal preliminarily was on its surface during formation.

Experiment and Discussion

By means of studying the morphology of domain structure of natural unipolar crystals and crystals with Cu^{2+} and Cr^{3+} dopants it has been determined that both the domain structure and the ferroelectric properties are homogenous along the growth layer of TGS crystals; their regular changes occur in direction of growth layers from the seed grain [2]. These results confirm the well-known thesis of A.V.Shubnikov about the polar direction being perpendicular to the growth pyramid.

The TGS crystals, grown under different conditions, have not strict correlation between static and dynamic unipolarity: the domain configurations with the close ratio of total squares of antiparallel domains S_+ and S_- can be repolarized to the quietly different (biased, double, with understated or with non-equal values of spontaneous polarizations P_s^+ and P_s^-) dielectric-hysteresis loops. Within single growth pyramid, one can see that single-domain state originates from the gradual displacement of one of domain components (Fig.1). In the growth pyramid {110} this process is followed with increasing the dislocation density, in growth pyramid (001) – with changing the geometry of domains, i.e. with the preferred orientation of domain walls by growth layers [3].

The establishment of mono-domain state is followed with the appearance and gradual increasing of intrinsic bias fields E_{bias} (Fig.2a) and their growth after establishment of mono-domain state.

* The unipolarity is characterized by electrostatic intensity E_{bias} in V/cm, biasing the center of dielectric-hysteresis loop to the axis of the fields.

In growth pyramid {110} the periodical changes of bias fields and dislocation density were also revealed (Fig.2b); $k_{bias}=E_{bias}/E_k$; E_k is the coercive field and σ is density of dislocations. Their morphological pictures corresponded to the domain configurations with the predomination of sign alternate spontaneous polarization P_s of domain components (Fig.2c).

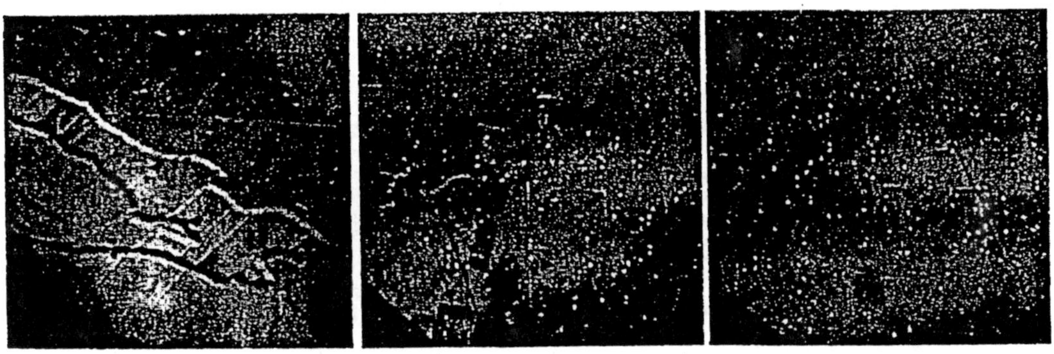

Figure 1 The change of domain configurations in the growth pyramid line {110} of TGS crystals.

In a strongly unipolar TGS crystals ($E_{bias}>E_k$) and TGS:Cu^{2+} crystals the correlation between E_{bias} and the length of helical dislocations in growth pyramid {110} was observed [2,3]. The helical dislocations are known to be the sink for the point defects and are resulted from interaction of point defects and screw dislocations. The "synchronous" increase of E_{bias} and length of helicoids is the indirect evidence of increasing the concentration of point defects with crysatliztion duration.

In poly- and mono-domain parts, P_s decreases, being more significant for larger E_{bias}.

There are considered individual examples of influencing the unipolarity upon some physical properties and behavior of domain structure under repolarization of TGS crystals.

In [4], the *P-T* diagrams of models of three types were investigated: the "ideal" model for $E_{bias}=0$; the naturally polarized specimen for $E_{bias}=125$ V·cm^{-1} and strongly unipolar specimen for $E_{bias}=-1400$ V·cm^{-1} doped with Cr^{3+}.

The phase *P-T* diagrams of naturally unipolar crystal and "ideal" crystal practically coincided: dt/dp for them is equal to 2.6 degree·kbar^{-1}. In a strongly unipolar specimen, dt/dp increased up to 3.1 degree·kbar^{-1}, the range of linear dependence of temperature of transition vs. pressure widened up to 8 kbar (Fig.3).

The inverse dielectric constant 1/ε appears to be "sensitive" to unopolarity itself: the Curie temperature and Curie-Weiss temperature (Δθ) did not coincide. This discrepancy (Δθ) depends on initial E_{bias} and increases with pressure, as well as the Curie-Weiss constant. For "ideal" specimen, 1/ε at phase transformation was constantly on the temperature axis up to 23 kbar.

The increase of E_{bias} with pressure can be evaluated, issuing from equivalence of effect upon dielectric characteristics of internal bias field and constant external field. For unipolar colemanite, Fatuzzo has shown [5] that internal field remains constant in range of ±5C with regard to transition point. Assuming that E_{bias} for the investigated TGS crystals submit to the same behaviors and using Vider's low

$$E=4/3[2A^3/3B]^{1/2}(\Delta T)^{3/2} \text{ and } (1/\varepsilon)_{max}=\pi A\Delta T, \qquad (1)$$

we obtain

$(1/\varepsilon)_{max}=0.3E^{2/3}B^{1/3}$. (2)

Assuming the coefficient B in first approach to be consent, we have

$(\varepsilon_{max})_2/(\varepsilon_{max})_1=[(E_{bias})_1/(E_{bias})_2]^{2/3}$. (3)

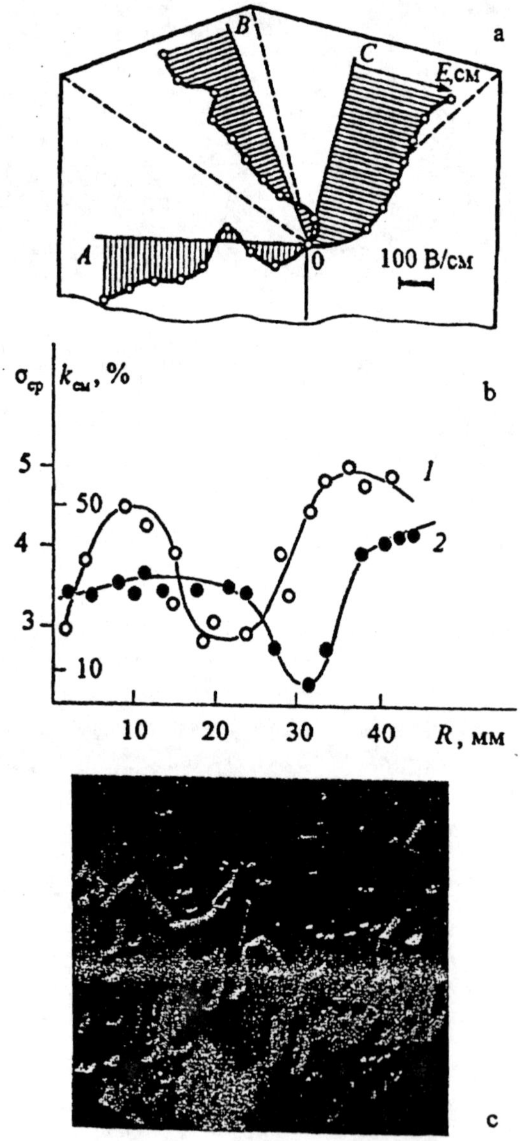

Figure 2 The change of domain configurations, the bias fields and dislocation densities in crystal growth line in TGS crystals: a – change of E_{bias} in OA, OC and OB in growth pyramid directions {110}, {001} and {121}, respectively (scheme); b – 1 – k_{bias} and 2 – lgσ in growth pyramid {110}; c – boundary (1) between domain configurations with predomination of P_s of different signs.

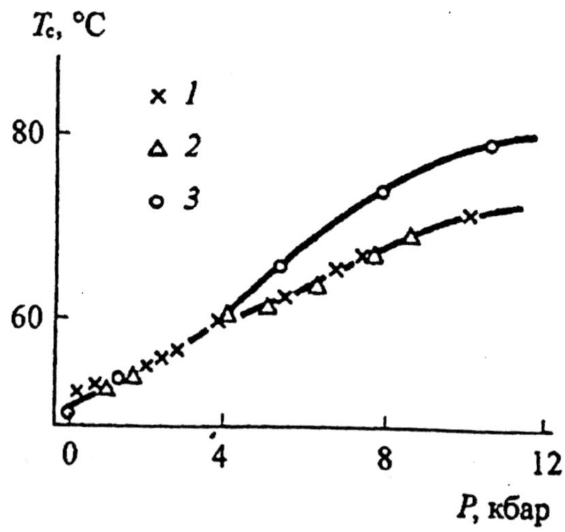

Figure 3 The phase P-T diagrams of ferroelectric transformation for TGS samples: for "ideal" (1), slightly (2) and strongly (3) unipolar crystals.

According to these evaluations, at pressure of 7 kbar the internal bias field increased for weakly and strongly unipolar specimens.

In unipolar crystals P_s=0 for temperature of ε maximum, the dependence P_s(T) in region of transition is blur, Curie temperature and Curie-Weiss temperature do not coincide. Such a characteristics, according to [6], makes indistinguishable phase transformation of 1st kind from the transformation of a higher order.

The natural unipolarity brings some difficulties at evaluating the thermal effects, occurring at phase transformation. Evaluating them by the jump of thermal capacity ΔC at transformation temperature, the jump, being $\alpha_\theta^2\theta$, must be equal to 13.8 cal·mole^{-1}deg^{-1} [7], provided that expansion of free energy occurs

$$\Phi = \Phi_0 + 1/2\alpha_0 T\text{-}\theta P^2 + 1/4\beta P^4 + \ldots \quad (4)$$

for $\alpha_0 = 3.92 \cdot 10^{-3}$ degree and $\beta = 8.0 \cdot 10^{-10}$ unit. SGSE.

The experimental value (Fig.4) exceeds the evaluated value already for two times as much. Besides, from expansion of free energy (4) it follows that there should not be the excess thermal capacity above transformation temperature. However, above transformation temperature it exists and in the range of order of 0.1C its gradually decreases (fig.4).

The appearance of excess thermal capacity above transformation temperature can be connected, in first, with "blurring" of phase transformation by internal bias fields E_{bias} [8] and, on second hand, with contributing the fluctuation polarization energy into free energy [9]. In the first case

$$\Delta C = \frac{\alpha_\theta'^2 T P^2}{\alpha_\theta'(T-\theta) + 3\beta P^2} \quad (5)$$

where P is defined from equation

$$\alpha_\theta (T-\theta)P + \beta P^3 = E, \quad (6)$$

one can calculate the influence of bias fields upon $\Delta C_E(T)$ and obtain the blurring of experimentally observed phase transformation.

In second case, according to [9], above to Curie point for one-axis ferroelectrics, the thermal capacity must depend on temperature with regard to logarithmic law. According to this theory, from the side of ferroelectric phase the thermal capacity must anomalously increase with nearing the Curie point. Such an increase was not observed experimentally (see Fig.4). This circumstance is obvious to be explained by the smallness of fluctuation contribution that anomaly is suppressed even by small bias fields which decrease with nearing T_c [4]. The heightened dispersion of thermal capacity near temperature of phase transformation is the evidence of the same fact.

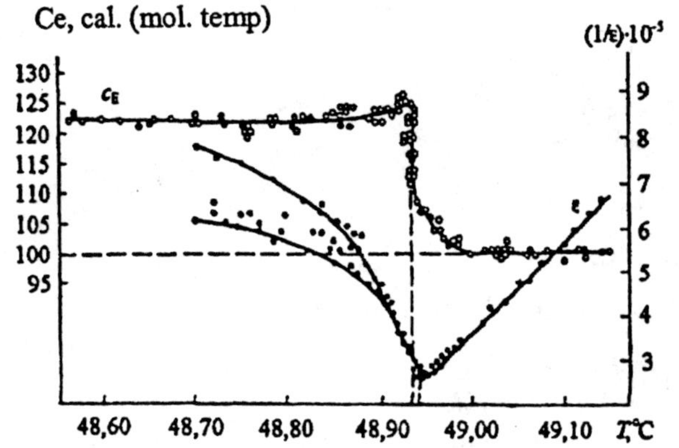

Figure 4 Thermal capacity and inverse dielectric constant of TGS at E=0.

As a rule, the position of experiment points of values of thermal capacity and 1/ε near phase transformation depended on time interval, separating the end of the annealing and the beginning of measuring the thermal capacity. In [7] this circumstance is connected with gradual appearance in polar phase of internal E_{bias} and their increasing with time, i.e. by soon ageing of the specimens with the characteristic time of 1-2 h. The annealing regime, which decreased but did not withdraw completely the natural unipolrity, was equal to 5 h at 90C. The ageing of crystal, caused by appearance of E_{bias} just after phase transformation, is reflected on dielectric constant behavior 1/ε. It follows the law of Curie-Weiss above the phase transformation temperature and lower only in range of T_c^{ε}-0.05C for lower temperatures. Below of the latter temperature, the dependence 1/ε(T) deflects from Curie-Weiss law, the values ε(T) are dispersed, which, obviously, reflect the process of enlargement of domain structure. By means of direct investigations of domain structure in range of 43-49C (including the phase transformation) the significant dispersion of average values of domain width due to [001] was revealed [10]. The reason of this dispersion is the regime of preliminary annealing (80C, 4 h), which is not sufficient neither by time nor by temperature for homogenizing the domain structure and decreasing E_{bias} down to zero, as shown in [11].

The beginning stage of repolarization studied in [12] by means of preferential etch method appeares to be dependent on E_{bias} in TGS specimens.

The preferential etch method is fixing the sequential positions of domain borders and allows observing the process of their isolation or the sealing on the dislocations.

If E_{bias}=0 and the walls are parallel to axis, i.e. are zero-charged, then the repolarization occurs due to the movement of walls of already-existed domains (Fig.5). If E_{bias}>>0, then the domain walls, inclined to the polar axis, are more inertness than for specimens with E_{bias}=0, and the repolarization begins from germ-formation process. The new domains usually appear near walls of "old" domains (Fig.6a), over the growth layers (Fig.6b), or they are localized in regions, optically unidentified to the morphology.

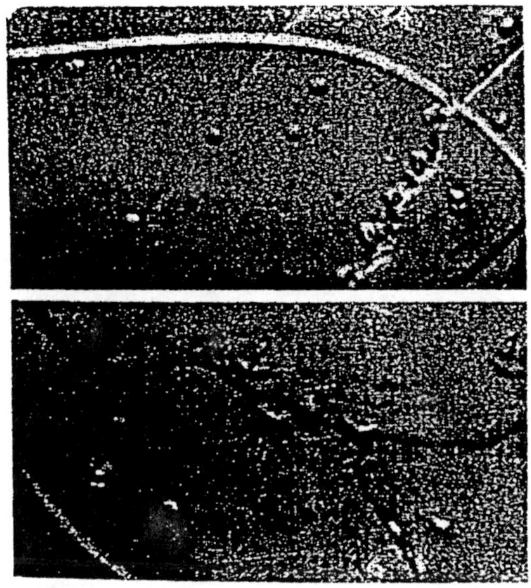

a

b

Figure 5 The beginning domain structure on the mirror chip (a) and sequence positions of domain boundary for "step" repolarization (b). One can see the catching of domain wall on the dislocations 1-4, its sagging due to the catching on the dislocations 1 and 2.

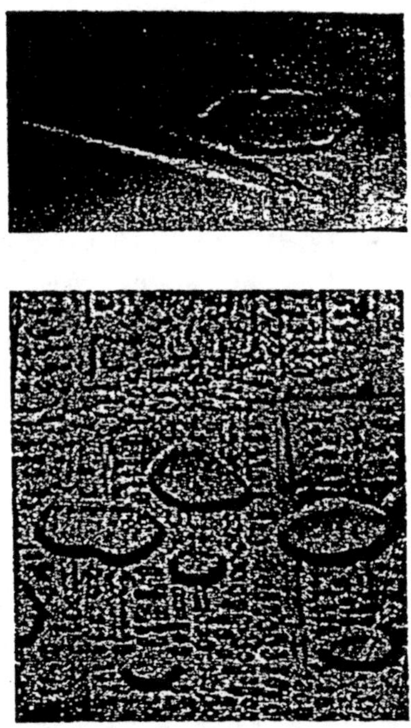

a

b

Figure 6 The nucleation of domains at perolarization of unipolar TGS specimens: a – the new domains appeared predominantly on perimeter of initial domain; b – the new domains of electro-thermal impact appeared along boundaries of growth layers; the boundaries of old domains are etched more deeply than the boundaries of new ones.

One can assume that one of the reasons of the stabilization of walls, inclined to polar axis, can be their heightened intrinsic energy, as far near these walls the bound-like charges are met. In order to compensate the latter, the free charges must flow out the bulk. The "clouds" of defect accumulations, scattering the light near walls of lens-like domain, were observed in [13], in fact.

For "developing" the places of localization of latent domain germs, the decoration method [14] appeared to be very successful one. This method is based on depending the rate of coalescence of decorating substance on charge surface sign and on the charge. The alter dependence allowed visualizing microscopic heterogeneities of charge distribution within one domain, i.e. inside domain and close to its boundaries [15].

The numerous observations of domain reconstructions in TGS crystals (under ageing, thermal impact, γ-exposure) are the evidences of that the growth layers often modulate the boundaries of a rebuilding domains (Fig.7a, b). In unipolar crystals and in the Cu^{2+} and VO^{2-} doped crystals there are a lot of heterogeneities and gradients of a dopant composition.** The etching polar chip structure of such crystals is the laminar one (Fig.8). The decoration reveals the growth layers of the same period as the preferential etching (Fig.9). Based on the decoration pictures and surface relief of the preferential etching, one can assert that within one domain, due to the layer variations of amount, there is modulated the value of spontaneous polarization itself. The analogous conclusions were made while studying the high-temperature ferroelectric $LiNbO_3$ [18]. The domain reconstructions at γ-exposure are various and also depend on the E_{bias}: the less latter, the more various initial domain configurations [19]. The role of energy anisotropy of domain walls in these processes has been revealed [20]. As was revealed, the increase of γ-radiation doze, the walls of orientations corresponding to the larger energies [102], [100], [101], [201], [001] include gradually into the orientations of the walls of new domains. According to [21], the following sequence of energies corresponds to them (the energy unit is the limiting energy for [001]): 0.58; 0.58; 0.67; 0.78; 0.8; 1.0. The direction [102] being one of low energies, greatly contributes into domain reconstruction, the new domains are lining up in {110} growth pyramid, appearing under γ-exposure, electron probe and thermal impact. The homogeneity of domain distribution, appeared under thermal impact, appears to be sensitive to gradient E_{bias} and was the larger, the smaller gradients.

The possibility of obtaining the time-stable and homogenous poly-domain state for the specimens with the initial unipolar structure and biasing fields was firstly demonstrated in [12], where the thermal and electric-thermal annealing was firstly demonstrated. It has been determined that timely irreversible decrease of E_{bias} starts only under annealing above some threshold temperature (110C), the time of whole annealing exponentially depends on temperature [2].

The specific conductivity σ_y in temperature range of 80-150C increases exponentially with temperature, the two branches differed by a abscissa axis slope in the linear dependence $\ln\sigma_y(1/T)$ are strongly expressed. The transition region is placed within the threshold temperature of annealing (110C); the $\ln\sigma_y(1/T)$ branch higher of 110C has the larger slope. The difference in the slopes, characterizing the energy of activation of process of irreversible annealing of natural unipolaity is 0.3 eV. The above-mentioned sharp curve in the dependence $\ln\sigma_y(1/T)$ disappears after finishing the annealing, when in ferroelectric phase the biased fields irreversibly decrease down to zero, and the domain structure becomes uniformly poly-domain.

** The investigations by means of resonance spectroscopy [16-17] show that dopants in structures of TGS crystals are placed without displacement of its individual elements, due to creation of complexes with them

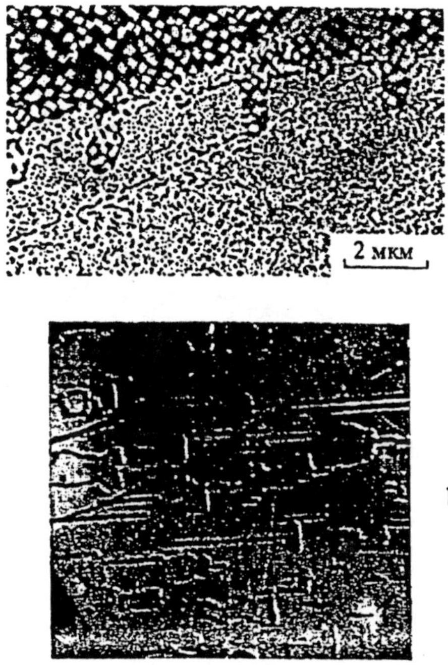

a

b

Figure 7 The formation of radiation domains in nonunitary TGS crystals: a – the dose of γ-radiation is equal to 3 mR; the thickness of decoration film d_{AgCl}=50 Å; b – the dose of γ-radiation is equal to 1.5 mR; the preferential etch (×100).

Figure 8 The domain and micro-zone structure of TGS:VO^{2-} crystal; the growth pyramid is {110}; the growth layers are parallel to [001].

The temperature of 110C was the highest among other characteristic temperatures revealed during investigation of decomposition process of TGS crystals [22]. The temperature structural instability of group NH_3, providing the radicals NH- and NH_2- [22], seems to be one of the reason causing natural unipolarity. It is difficult to free from this unipolarity even by means of up-to-date purification methods of primary products of TGS crystals [23].

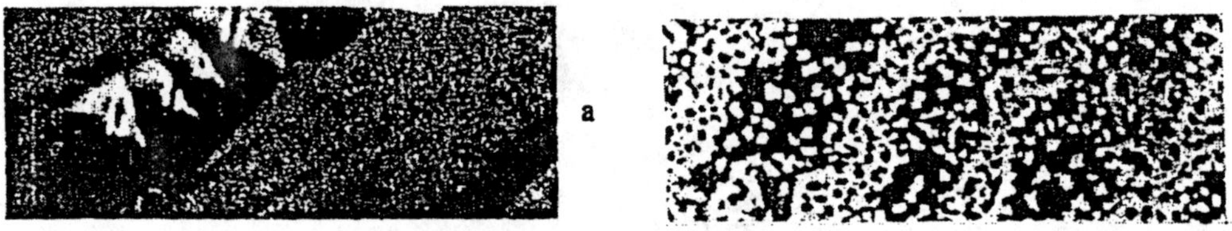

Figure 9 The banded structure of domains, "developed" by γ-radiation (15 mR) of TGS crystal; the thickness of decoration film d_{AgCl}=50 Å.

Taking into consideration the thermal instability of glycine groups, one can assume that at TGS crystal growth, the fragments of groups in a pure appearance or the bounded with "background" dopants (Si, Ca, Mg), already are those (thermally destroyed) point defects, which take part in formation of helical dislocations, the density of which changes synchronously to E_{bias}.

Comparatively small activation energy necessary for irreversible annealing of natural unipolarity (0.3-0.4 eV), allows assuming that it is spent for small changes in the position of dopant complexes (or in their structure) under temperature and/or electric field. The evidences of disorder in doped and intrinsic structure of TGS crystals is the disappearance of micromicro-zonary dark streak pattern after annealing of naturally-unipolar specimens [2].

Ad hoc introduced paramagnetic dopants (Cu^{2+}, Cr^{3+}, VO^{2-}) create stable complexes with the element of structure of TGS glycine-I that aloows using them as structural sondes for studying the phase transformations and repolarization processes in TGS crystals by means of electron spin resonance [16-17]. The high-temperature annealing is unable to destroy these complexes irreversibly, therefore there are no anomalies $\sigma_y(T)$ in region of 110C.

As it was mentioned above, the stable mono-domain state is the condition for a lot of technically important usages of TGS crystals. All the attempts of its obtaining with the help of dopants or α-alanine, according to our experience, was not successful, as well as the experiments on mono-domainization of specimens at temporal effect of strong direct current and γ-radiation: in TGS specimens the domains which did not obey to time-independent electrical field had the total area from 0.1 up to 10 rel.% for the dose of γ-radiation of $1<D<5$ mR [15].

In the light of above-stated, it is clear that the reason of incomplete mono-domainization of specimens under any effects is their biographical and growth heterogeneity caused by non-stationary crystallization regimes. These deviations develop in the different characteristic periods substructures: in the micro-zonal of period of 1-5 μm (determined from the pictures of zonal melting [2,12]); in the macro-periodical structures with the period of oscillation of several mm (determined from characteristic changes of domain structure and variations of E_{bias}).

The TGS crystal "forgot" its biography and obtained stable and homogenous state in its domain structure (even at electron-microscopic level [19]) due to three effects: 1 – a high-temperature annealing; 2 – γ-radiation (the dose of 15 mR) at temporally applied electric constant field (3) (for order as much of coercive field) [24].

The crystals of technical importance, including $LiNbO_3$, are basically grown from congruous compositions corresponding to instable components [25-26]. As the local investigations of $LiNbO_3:Fe^{3+}$ have shown (Fig.10), this instability develops as the chaotically periodic variations of composition (Fig.10b). Only individual parts of trajectories of phase portrait have a strongly expressed periodicity (Fig. 10d), as though striving to one utmost cycle, but chaotically alternating it [27].

In single crystals $LiNbO_3:Fe^{3+}$ (3 mass.%), the variation of amount for intrinsic (Nb, oxygen) and alloying (Fe) elements were studied by means of local X-ray microanalysis method (LXRMA). The distribution of Nb and Fe, regarding to the domain walls, have allowed determining the picture of preferential etch of domains (and dislocations) and the traces of snuff of electron sonde (Fig.10a). As seen from comparison of Fig.10b and Fig.10c, the extreme character in distribution of Fe is for the walls of both types, but for walls P_s^+/P_s^+ these values are minimal, and for walls P_s^-/P_s^- is maximum with regard to the neighboring compositions. The same tendency is found for variation of Nb amount.

We assume that composition modulation for alloying (Fe) and basis (Nb) components allows understanding the appearance of domain walls being perpendicular to P_s at paraelectric-ferroelectric phase transformation in lithium niobate (if crystal is not undertaken to mono-domainization with the help of constant electrical field while growing). For the wall of P_s^+/P_s^+ type in the moment of ferroelectric phase transformation it is "convenient" to be placed where it will be surrounded by negatively charged and incompletely point defects. In our case Fe and Nb vacancies can be those point defects (the violation of flat Nb infill in $LiNbO_3:Fe^{3+}$ were revealed in [18] by means of high-resolution electron microscopy).

Investigation by means of LRSMA has revealed the macroscopic ~0.5 mm variation period of Nb amount (Fig. 10b). The nature of macroscopic variations of crystal compositions, appeared during both the growth from solution growth and the melt growth, requires the studies. The nature of this process is based on physic-chemical processes [28], passing in growth systems on the boundaries of solid and liquid phases.

A more precise determination of congruous composition for $LiNbO_3$ has been carried for 20 years. The results of work [29] seem to be most conclusive, where the change step of stoichiometry of initial melt was 0.05 mol.%, and the stoichiometry of the obtained crystals was determined due to temperature dependence of phase ferroelectric transformation and phase concordance temperature for non-critical generation of quadratic component. The investigations of domain structure have shown that while passing through congruous composition (48.38±0.015 mol.% of Li_2O [29]) the predominant domain structure changes the sign, and the structure of crystal in congruous point is rigorously poly-domain.

In TGS crystals the analogous process of evolution of domain structure is extended, the initial poly-domain structure transfers gradually into mono-domain one (Fig.1); in the individual cases the periodical change of predominant domain component and a sign of bias fields is observed (Fig.2c). The latter, because of a domain structure and because of practical impossibility of controlling the local (for lithium niobate) stoichiometry of TGS crystals, can be considered as the indicator of stoichiometric "prosperity" or TGS crystallization process.

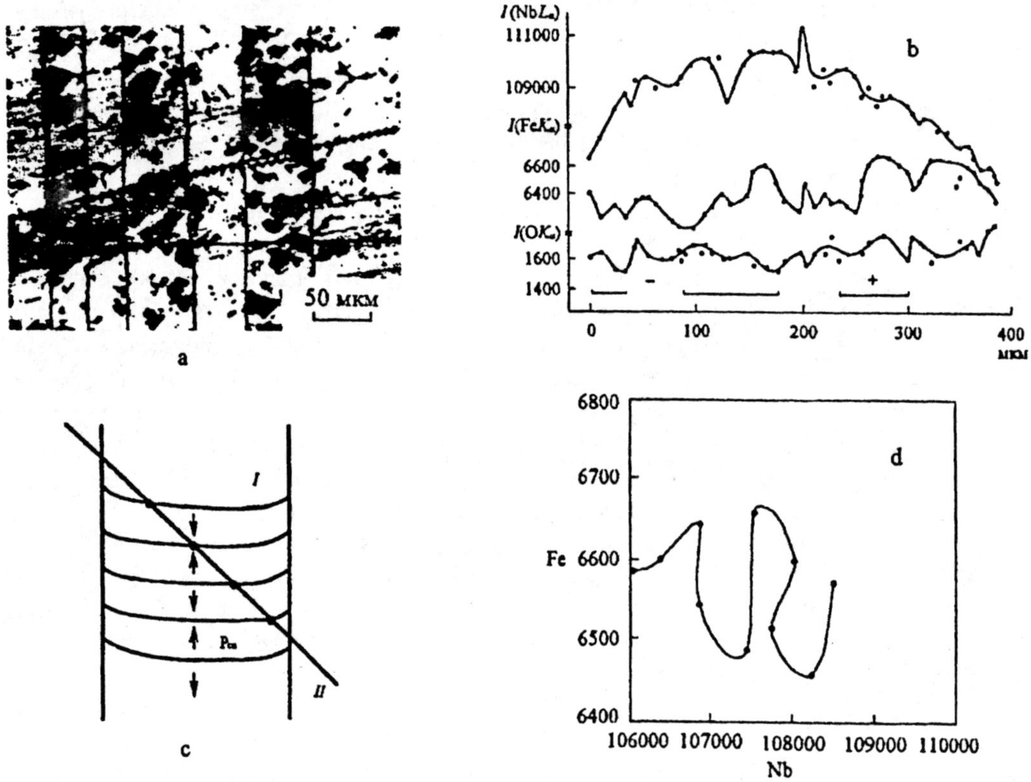

Figure 10 The domain structure, dislocations and variations of composition of crysatal $LiNbO_3$:Nb^{3+} (3 mass.%) with Nb, Fe and oxygen: a – the domain structure, dislocations and traces of electron sonde on surface of chip being rhombohedron (102) of crystal; b – distribution of intensity of Nb L_α Fe K_α and O K_α, measured in local parts (see a) by means of local X-ray micro-analysis method (microanalyser "Camebax"); c – the scheme of position of chip-etch plate (II) and the layers of domains (I); d – the part of phase portrait in co-ordinates Nb-Fe (according to data of 10b.)

Conclusion

The physic-chemical state of initial medium (the melt stoichiometry in the case of lithium sniobate, the pH in the case of solution and the amount of dopants in the case of triglycinesulphate) and the physic-chemical state of formed crystals are connected with some periodical processes. Their appearances are revealed by means of different resolution methods and have different periods: ~5 μm (the growth layers, the preferential etch and the electron-microscopic decoding, TGS); 50-70 μm and 400-500 μm (composition variation in intrinsic and alloying element, LXSMA method, $LiNbO_3$:Fe^{3+}); several mm and even cm (variation of amlitude and sign of E_{bias}, with respect to dielectric hysteresis loops).

Comparing the results obtained at study of domain structure: *a* – along direction of TGS crystal growth (in a given work) and *b* – for variation of stoichiometry of initial melt near congruous composition in lithium niobate crystals [29], it follows that for water soluble crystals the bias fields (accompanying the variations of domain structure) are the fine and, obviously, the only indicator of change of crystallization conditions. To the latter, one should relate not only the concentration of intrinsic and alloying elements (and "background" dopants), but also their mutual, "phase" [26] changes.

The bias fields change the physical properties, in particular, a thermal capacity under ferroelectric-paraelectric phase transformation, suppressing the fluctuations of spontaneous polarization in transformation region.

The composition variation in the intrinsic and alloying (Nb and Fe) components, revealed by LXMA method, have allowed explaining the anomalous localization of charged domain walls in crystals $LiNbO_3:Fe^{3+}$.

Based on the investigation of domain structure, temperature of bias fields and conductivity of TGS crystals, the complex of physical treatments (high-temperature annealing followed with γ-radiation, under applying the electric field in direction of former preferential domain orientation) allowing obtaining the homogenous and a stabilized mono-domain state being necessary for using the TGS crystals as the high-sensitive pyro-receivers have been observed.

We suppose that the understanding and the know-how "reading" of morphologic substructures of ferroelectric crystals is necessary for expert scientific interpretation of obtained experimental results (the investigations of their very structurally sensitive properties) and for express-evaluation of availability of their practical usage.

REFERENCES

1. Lemmleyn G.G. Morfologiya i genezis kristallov. M.:Nauka. 1973, 327 s.
2. Meleshina V.A. Domennaya struktura i unipolyarnost kristallov triglitsinsulfata: Dis.kand. fiz.-mat. nauk. M., 1966, 180 s.
3. Koldobskaya M.F., Meleshina V.A., Rez I.S. i dr. Kristallografiya, 1966, t. 11, № 5, s. 785-793.
4. Leonidova GG., Netesova R.P., Meleshina V.A. i dr. Vestnik MGU, Ssr. 4, Geologiya. 1971, № 4, s. 436-439.
5. Fatuzzo E. J.Appl. Phys., 1960, v. 31, no. 6, p. 1029-1034.
6. Asoulay J., Gringberg J., Pelah J. et al. J.Phys.and Chem. Solids., 1968, v. 29, no. 5, p. 843-849.
7. Taraskin S.A., Strukov B.A., Meleshina V.A. Izv. AN SSSR. Ser.fiz, 1965, t. 29, № 6, s. 982-984.
8. Strukov B.A., Taraskin S.A., Skomorohova T.L. i dr. Izv. AN SSSR. Ser.fiz., 1965, t. 29, № 6, s. 982-984.
9. Levanyuk A.P. Izv. AN SSSR. Ser.fiz., 1965, t. 29, № 6, s. 879-881.
10. Strukov B.A., Meleshina V.A., Kalinin V.I. i dr. Kristallografiya, 1972, t. 17, № 6, s. 557-562.
11. Meleshina V.A., Rez I.S. Izv.AN SSSR. Ser. fiz., 1964, t. 28, № 4, s. 735-740.
12. Meleshina V.A. Kristallografiya, 1971, t. 16, № 3, s. 557-562.
13. Taurel L., Gilleta F. Proc. of the Intern. Meet. on Ferroelectriciy. Prague, 1965, v. 2, p. 43-50.
14. Vlasov V.P., Gerasimov Yu.M., Distler G.I. Kristallografiya, 1970, t. 15, s. 346.351.
15. Belugina N.V. Domennaya struktura, neodnorodnost polyarizatsii i nekotorye fizicheskie svoystva kristallov TGS s razlichnoy stepenyu defektnosti. Dis. kavd.fiz.-mag. nauk. M., 1977, 210 s.
16. Waplak S., Yurin V.A., Stankovski et al. Acta Phys. Pol. A, 1974, v. 46, p. 175-183.
17. Windsch W., Volkel G. Ferroelectrics, 1978, v. 17, no. 3/4, p. 491-500.

18. Meleshina V.A., Bagdasarov Ch.S. Minjukov S.A. et al. Phys. Status Solidi A, 1980, v. 62, p. 475-482.
19. Yurin V.A., Belugina N.V., Meleshina V.A. i dr. IZV.AN SSSR. Ser.fiz., 1971, t. 35, № 9, s. 1927-1930.
20. Meleshina V.A., Belugina N.V., Shlemmbah X., Vindsh V., Gavrish L. Kristallografiya, 1995, t. 40, s. 863-867.
21. Hatano J., Suda P., Futama H. Ferroelectrics, 1978, v. 20/22, p. 265-266.
22. Ziebert V. Ferroelectrics, 1974, v. 6, no. 3/4, p. 157-162.
23. Malec Z., Moravec F., Strajblova J. Chechosl. J. Phys., 1969, v. 19, no. 9, p. 1184-1185.
24. Belugina N.V., Meleshina V.A., Yurin V.A. i dr. Kristallografiya, 1975, t. 20, № 2, s. 438-440.
25. Errousmit D., Pleys K. Obyknovennye differentsialnye uravneniya. M.: Mir, 1986, 244 s.
26. Meleshina V.A., Fedorov P.P. Doklady RAN, 1998, Ser. Fizicheskaya himiya, 1999, № 6, (v pechati).
27. Prigozhiy I., Stengers I. Poryadok iz haosa. M.: Progress, 1986, 431 s.
28. Shemyakin F.M., Mihalev P.F. Fiziko-himicheskie periodicheskie protsessy. Moskva-Leningrad: AN SSSR, 1938, 183 s.
29. Bordui P.P., Norwood R.G., Bird C.D., Calvert G.D. J.Cryst. Growth., 1991, v. 113, no. 1, p. 61-68.

About the Authors

Valentina Alexandrovna Meleshina *– The Institute of Crystallography of RAS, Candidate of Physics and Mathematics. A Research Officer. An expert in area of investigation of morphology and genesis of crystals by means of preferential etch methods, optical and scanning electron microscopy and X-ray microanalysis.*

Natalia Vasilievna Belugin *– The Institute of Crystallography of RAS. A Candidate of Physics and Mathematics, Research Officer. An expert in area of investigation of morphology, domain real structure of ferroelectrics by means of preferential etch methods, electron-microscopic decoration and X-ray microanalysis.*

GROWING CR^{4+}: CA_2GEO_4 LARGE CRYSTALS BY SOLUTION-MELT METHOD

V.V. Dozmarov, L.D. Iskhakova, V.N. Protopopov and E.M. Dianov

This work is devoted to searching the solvents that enable the use of ordinary solution-melt method for producing Ca_2GeO_4 single crystals (these conditions are: the Pt-crucible, a low (to 1250^{0}C) temperatures, the resistance heating). The series fluoride (NaF, MgF_2, CaF_2 and PbF_2) as well as series oxides (B_2O_3, P_2O_5, (Li_2O+MoO_3)) have been investigated. The solvent based on Li_2MoO_4 appears to be the most promising for growing the crystals Ca_2GeO_4. There was used the top-seeded solution growth technique in combination with pulling the crystal. The dimension of crystals obtained by a given method is 60×50×30 mm, and the weight is 100 g.

INTRODUCTION

Presently, the reasearches on tunable lasers based on crystals in near IR-range (1.2-1.5 μm) are intensively developing. Demonstrated in 1988 [1], forsterite (Cr^{4+}:Mg_2SiO_4) laser had raised the interest to this material. Since that time, the active search on new perspective laser single crystals among the silicates and germinates with Cr^{4+} ions in tetragonal co-ordination [2.3] had started. But the problem of creation of the high efficiency and the low excitation threshold active laser mediums for solid-state tunable lasers in the near IR-range is not solved yet. This is connected, first of all, with weak luminescence or its absence because of non-radiating in series materials. It is difficult to obtain single crystals of the intensive luminescence, the high optical properties and the necessary dimensions.

The single crystals Ca_2GeO_4 are among the most promising materials [3] for solving this problem. The first attempt to grow crystals and determine the structure Ca_2GeO_4 was made in 1948 [4]. The phase diagram of system CaO-GeO_2 has been investigated in 1966 [5]. It has been revealed that Ca_2GeO_4 has two polymorphous modifications with temperature of phase transition of 1453^{0}C. The low-temperature phase is isostructural one to the forsterite Mg_2SiO_4, possessing the structural type of the peridot (the *Pbmn* space group [6]).

Up to present time, the most significant results on crystals growth and practical use of Cr^{4+}:Ca_2GeO_4 are presented in [3], authors of which have grown Cr^{4+}:Ca_2GeO_4 single crystals by modified Chokhralskyy method from the solution in the melt at temperature being near the temperature of phase transition, under special conditions (induction heating, iridium crucibles, nitrogen). The grown single crystals were of 15 mm in diameter and 30 mm in the length.

GROWING Ca_2GeO_4 CRYSTALS

The given work is devoted to searching the solvents for growing the Cr^{4+}:Ca_2GeO_4 single crystals by ordinary solution-melt method, allowing operating in range of low temperatures (down to 1250C, the platinum crucibles, resistance heating). To this end, the crystallization of Ca_2GeO_4 in series of oxides and oxide-fluoride systems CaO-GeO_2-L, where L=B_2O_3, V_2O_5, P_2O_5, (Li_2O+MoO_3), NaF, MgF_2, CaF_2 and PbF_2 has been studied.

The technique investigation of phase formation and determination of fields of crystallization of Ca_2GeO_4 came to carrying out the trial crystallizations in small volumes of melt (~1 cm^3) for different molar ratios of components. The experiments were carried out from the melts near pseudo-binary crosscut of Ca_2GeO_4-L, where the amount of L was measured in range 20-50 moles.%. The temperature of beginning of crystallization was evaluated from precipitating the spontaneously formed crystals on the platinum wire-probe placed into the solution and periodically drawn-out due to decreasing the temperature for microscope visual control. The cooling rate of the system in such experiments was 20-50C/h. The germanium dioxide of pureness of 99.999 %, the extra-pure Li_2CO_3, B_2O_3, NaF, PbF_2 and V_2O_5, $CaCO_3$ and CaF_2 were used as the primary powders.

The identification of the crystallized phases was carried out with the help of X-ray analysis. The shooting of specimens was carried out automatically on the diffractometer DRON-4-13 (CuK_α-radiation, germanium or silicon of semiconductor pureness as the standard). For processing the experimental results and carrying out the X-ray analysis, the "RAPID" software was used. Indication of X-ray pictures, the more accurate determination of parameters of unit cells of compounds and calculation of theoretical X-ray pictures due to the given X-ray data was carried out with the help of software "POWTOOL". The determination of X-ray characteristics of calcium fluorogermanates and more precise determination parameters of their unit cells were carried out on the four-circular monocrystal diffractometer CAD4 with Mo-radiation (the shooting was performed by A.V.Yatsenko, MSU).

The morphology of crystals was studied with X-ray methods using plant for orienting the large dimension crystals on the base of DRON-2 with attachment PRF-1 and diffractometers URD-63 and DRON-4-13 for work with crystals of less dimension.

The oxides of phosphorus, vanadium and boron are not perspective for growing Ca_2GeO_4 crystals. The phase-formation in systems with two first oxides leads to crystallization of $Ca_2P_2O_7$ or vanadates of calcium, respectively. In the boron oxide rather than required rhombic shape of Ca_2GeO_4, having the structure of peridot, the hexagonal modification of Ca_2GeO_4 is stabilized by boron oxide (the cell parameters: a=5.514(2) Å, c=7.160(2) Å, the space group is $P6_3mc$) and its isolation from the melt occurs.

The crystallization of Ca_2GeO_4 in the sodium, magnesium and calcium fluorides is complicated by formation of fluorogermanates. The parameters of unit cells of these compounds, crystallized in calcium and sodium fluorides, are given in Table.

The comparison of obtained results to structural data for compounds of germinates has shown that X-ray characteristics of phases #1, 2 and 3 are analogous to those for structure of fluoride-placed calcium germanate of compound $Ca_{3-x}[GeO_4](O_{1-2x},F_{2x})$ of x=0.275, obtained by authors [7]. The parameters of unit cells of phases #4-7 allow interpreting them as the superstructure disproportionate modulated phases [8] with respect to primary pro-structure of $Ca_{3-x}[GeO_4](O_{1-2x},F_{2x})$. The comparison of cell parameters of the obtained phases to the parameter of cell of fluoride-germanate with x=0.275 allows concluding the variation of fluoride amount in a given compounds.

Table: The Parameters of Low-Level Cells of Fluoro Germinates

The phase	*a*,	*c*,	Space group
$Ca_{3-x}[GeO_4](O_{1-2x}F_{2x})$, $x = 0.275$ [7]	$a_0 = 7.223$ (1)	$c_0 = 5.681$ (1)	$P6_3mc$
№ 1, system with CaF_2	7.203 (3)	5.760 (1)	$P6_3mc$
№ 2, system with CaF_2	7.202 (2)	5.735 (1)	$P6_3mc$
№3, system with CaF_2	7.198 (2)	5.755 (1)	$P6_3mc$
№4, system with CaF_2	$a \cong 5a_0 = 36.002$ (3)	$c \cong 3c_0 = 17.273$ (9)	
№ 5, system with CaF_2	$a \cong 2a_0 = 14.637$ (3)	$c \cong c_0 = 5.389$ (2)	P3
№ 6, system with NaF	$a \cong 2a_0 = 14.605$ (3)	$c \cong c_0 = 5.388$ (5)	P3
№ 7, system with NaF	$a \cong 2a_0 = 14.645$ (3)	$c \cong c_0 = 5.398$ (6)	P3

In the lead fluoride based system, the X-ray analysis of the specimens has revealed the formation of PbO and lead germinates as the phases, accompanying the crystallization of Ca_2GeO_4.

Such a traditional compounds as halogenides were studied as the solvent. The bulk crystallization of rhombic calcium germanate occurs in medium CaO-GeO_2-$CaCl_2$. Because of high volatility of calcium chloride (during 24 hours the mixture mass loss was about 15%) and the complexity of treating the hygroscopic calcium chloride (while preparing the mixture, there was fixed the appearance of CaOCl phase), a given system is unacceptable for growing the single crystals of required quality.

The investigations, carried out in the Li_2MoO_4 based solvent, have shown that given system is perspective for growing the crystals Ca_2GeO_4. In combination with Ca_2GeO_4-Li_2MoO_4, the calcium germanate has the lengthy crystallization area; the advantage of these solutions-melts is their small viscosity and insignificant volatility of the components. The parameters of Ca_2GeO_4 rhombic unit cell are: a=5.226(1) Å, b=11.401(2) Å, c=6.750(1) Å, the space group is *Pbnm*.

For obtaining the single crystals Cr^{4+}:Ca_2GeO_4, the top-seeded solution growth method has been used in combination with drawing the crystal. The crystal growth plant is the resistance-heating furnace. The accuracy of temperature regulation was ±0.05C. The dimension of crystals obtained by a given method achieves 60×50×30 mm, and the weight is 100 g, Fig.1. The parameters of unit cell of chrome activated single crystals are: a=5.2420(5) Å, b=11.4002(9) Å, c=6.7773(7) Å.

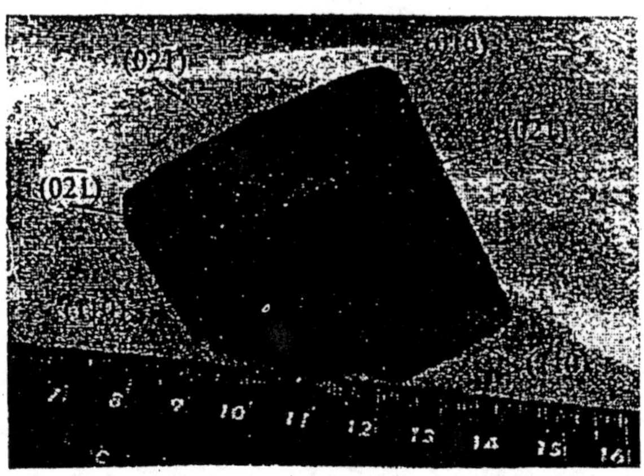

Figure 1. The picture of Ca_2GeO_4 single crystal

The shape of crystals Ca_2GeO_4 depends on the orientation of seed. The crystal illustrated in Fig. 1, has been grown using the prism seed, cut perpendicular to a plane (110). It is seen from fig.1 that morphology of crystal is simple and is determined by {110} facet of prism (mostly developed), as well as by {021} and {101} facets of prism. The facets of side pinacoid {010} are narrow. By using the seed, cut along [010] direction, the facets {021} of prism are mostly developed; the facets {110} of prism are essentially small; the facets {010} of pinacoid are very narrow ones (almost degenerated).

SPECTROSCOPIC INVESTIGATIONS

For spectroscopic investigations the oriented cube of dimension 5×5×5 mm, cut along crystallographic from grown boule, has been used. The accuracy of orientation of specimen was controlled on the X-ray diffractomeetr, and was not worse of 2^0.

The absorption spectra of crystals Ca_2GeO_4, doped with chrome, were recorded on the spectrometer Perkin Elmer Lambda-900 and are presented in fig.2.

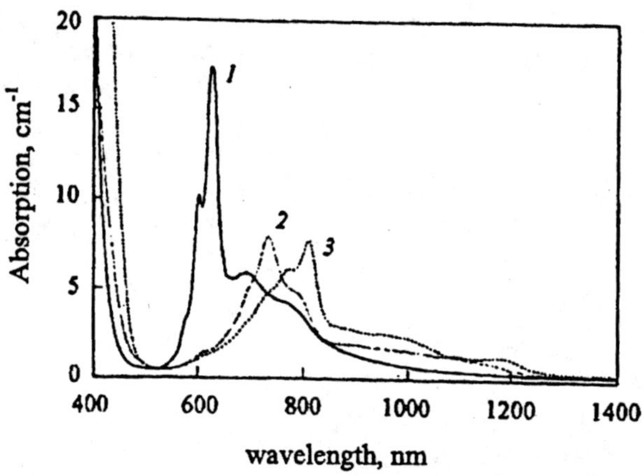

Figure 2. The absorption spectra of Cr^{4+}:Ca_2GeO_4 for three different orientations. 1 – E||a; 2 – E||c; 3 – E||b.

The wide intensive bands with peaks near 630, 740 and 820 nm correspond to transition $^3A_2$ $^3T_1$ of Cr^{4+} ion in tetragonal crystalline field. The wide absorption bands in IR-range (900-1200 nm) correspond to transition $^3A_2$ $^3T_2$ of this ion.

The calcium germanate, doped with chrome, excited by radiation of argon laser (λ=488 nm), demonstrates the intensive luminescence in the nearest IR-region. The investigation of spectra was carried out at room temperature. This luminescence is caused by radiative transition $^3T_2$ $^3A_2$ of ions Cr^{4+}. The luminescence spectrum (Fig.3) is the wide band in range 1150-1550 nm with peak in region of 1270 nm. In contrast to forsterite doped with chrome, in Ca_2GeO_4 luminescence spectrum, the bands corresponding to Cr^{3+} are absent.

From the analysis of luminescence decay kinetics at room temperature, the time of dissociation of excited level $^3T_2$ of ions Cr^{4+}, being equal to ~12 μs, exceeding for several times the forsterite's one, has been obtained. Such a significant decrease of lifetime of excited state can be the evidence of decreasing the probability of non-radiating poly-background relaxation. The weak dependence of lifetime of excited state vs. temperature (fig.4) in comparison to forsterite is indirect proof of the latter. The chrome concentration in Ca_2GeO_4, being equal to 0.05 mol.%, does not cause the quenching with concentration.

Evidently, that decrease of probability of irradiating poly-background relaxation must lead to increasing the quantum yield of luminescence.

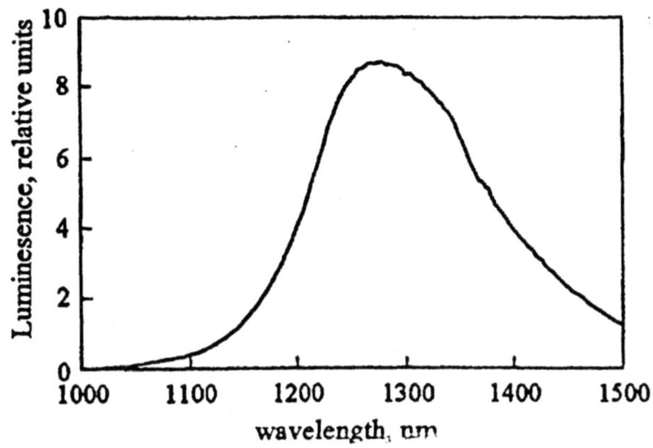

Figure 3. Luminescence spectrum of Cr^{4+}:Ca_2GeO_4 crystal.

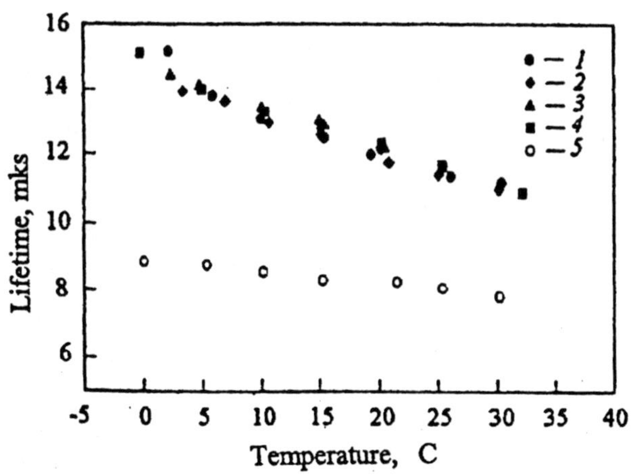

Figure 4. Temperature dependence of life-time of excited state Cr^{4+} in different specimens of Ce_2GeO_4. 1 – the specimen #1, 0.017 mol.% Cr; 2 – the specimen #2, 0.032 mol.% Cr; 3 – specimen #3, 0.051 mol.% Cr; 4 – specimen #3c, 0.051 mol.% Cr, 5 – the specimen B, 0.618 mol.% Cr.

It is expected that improvement of growth technology will allow producing the Ca_2GeO_4 single crystals with chrome of laser quality, in which the concentration of chrome in Ca_2GeO_4 will significantly exceed the concentration of chrome in forsterite that will allow creating tunable lasers with low lasing-action threshold and high efficiency, operating at room temperatures.

CONCLUSIONS

1. Resulting from searching the solvent, the solvent on the base of Li_2MoO_4 was determined to be the perspective system for growing the Ca_2GeO_4 crystals.
2. It is first, the single crystals Cr^{4+}:Ca_2GeO_4 have been grown by ordinary solution-melt method (temperature up to 1250C, the platinum crucibles, the resistance heating). The dimension of crystals obtained by a given method is of 60×50×30 mm and of weight up to 100 g.
3. It is first, the morphology of a grown crystals Ca_2GeO_4 was described and the facets were indexed.

The work has been performed for financial support of RFFI (grand #98-02-17340).

REFERENCES

1. Petrecevic V., Gayen S.K., Alfano R.R., Yamagishi K., Anzai H., Yamaguchi Y. Appl. Phys. Lett., 1988, v. 52, p. 1040-1045.
2. Borodin N.I., Zhitnyuk V.A., Ohrimchuk A.G, Shestakov A.V. Izv. AN SSSR. Ser. Fizicheskaya, 1990, t. 54, № 8, s. 1500-1506.
3. Petrecevic V., Bykov A.V., Evans J.M., Alfano R.R. Opt. Lett., 1996, v. 21, p. 1750-1752.
4. Zhuravlev V.F. Dokl. AN SSSR, 1948, t. 59, s. 1145-1148.
5. Shirvinskaya A.K., Grebenschikov R.G, Toropov N.A. Izv. AN SSSR. Neorg. mater., 1966, № 2, s. 332-336.
6. Belov N.V., Belova E.N., Andrianova N.N., Smirnova R.F. Dokl. AN SSSR, 1952, № 3, s. 81-84.
7. Nishi K,, Takeuchi Y. Acta Cryst., 1984, C 40, p. 730-733.
8. Strukov B.A., Levanyuk A.P. Fizicheskie osnovy segnetoelektricheskih yavleniy v kristallah. 1995, M.: Nauka, 240 s.

ABOUT THE AUTHORS

***Vitaliy Vital'evich Dozmarov** – the Scientific Center of Fiber Optics of Institute of General Physics of RAS, a research fellow. An expert in area of the high-temperature melt and solution crystals growth.*

***Ludmila Dmitrievna Iskhakova** – the Scientific Center of Fiber Optics of Institute of General Physics of RAS, a candidate in Chemistry, a research officer. An expert in area of crystal chemistry and non-organic chemistry.*

***Vladimir Nikolaevich Protopopov** – the Scientific Center of Fiber Optics of Institute of General Physics of RAS, a research fellow. An expert in area of the optics and spectroscopy.*

***Evgeniy Mikhaylovich Dianov** – the Scientific Center of Fiber Optics of Institute of General Physics of RAS. A Doctor in Physics and Mathematics, a Member of the Academy, director. An expert in area of fiber optics.*

Valeriy Veniaminovich Voronov *– the Scientific Center of Fiber Optics of Institute of General Physics of RAS, a Candidate in Physics and Mathematics. A Chief of Sector of IOF RAS. An expert in area of solid-state physics.*

Oleg Konstantinovich Mel'nikov *– the Institute of Crystallography RAS, a Candidate of Chemistry, a research officer. An expert in area of growing the crystals.*

CARBON-GLASS CERAMICS AND ITS PROPERTIES

V.F. Tatarinov, Yu.S. Virgil'ev and S.V. Evdokimov

The work summarizes the published data on the properties of carbon-glass ceramics, developed early and obtained recently by modified technology. We present a statistical distribution of some characteristics of industrial carbon-glass ceramics used for producing the artificial heart valves.

INTRODUCTION

The carbonic structural materials are widely used in different areas of science and engineering. The carbon-glass ceramics is used in medicine owing to the high constructional and fatigue strength, the wear resistance for the lifetime, the good machinabillty, the impermeability with respect to liquids and the biological compatibility with blood. The carbon-glass ceramics itself and the products of its consumption are not toxic. Such an optimal combination of properties has allowed producing the actuated parts of artificial heart valves [1].

The aim of this work is to summarize the collected data on physical properties of carbon-glass ceramics concerning its usage as the actuated part of artificial heart valve. The given material can be used in other areas of science and engineering.

OBTAINING THE MATERIAL

The carbon-glass ceramics possess to the isotropic thermal carbon materials of condensation-crystallization structure [2]. This material is created during deposition onto reaction surface of carbon disperse particles, formed in the gas phase and being bound by pyrolytical carbon [3]. The number of such particles (globules) achieving the certain part of reaction surface is determined by the gas-dynamic conditions at dust-laden flow streamlining the irregularities of graphite substrate in reactor. Depending on it, the concentration of the disperse particles and the permolecular structure of material can be varied from pseudo-isotropic to the crystal-oriented one. The ratio of the disperse particles number and the binding pyrolytic carbon determines the properties of the obtained material [4].

Presently, with one of the technologies, USB-15 carbon-glass ceramics is obtained in tubular axially symmetric reactor of 120 mm in diameter at temperature of 1500C by means of oriented crystallization of carbon at pyrolysis with the boron trichlorid catalyst. The carbon-glass ceramics lies onto internal surfaces of flat substrates of 40 mm in width, assembled into polyhedron of 80 mm in diameter [5].

For such a technology layering the carbon-glass ceramics at the substrate center and on the edges goes for different rate. The latter also grows to the end of the process because of decreasing the cross-section of reaction zone owing to layer growth, layered up to a given thickness (3-10 mm) of the carbon-glass ceramics. Besides, the adhesion of carbon-glass ceramics to the MPG graphite having a higher thermal coefficient of linear expansion (TCLE) in comparison to the carbon-glass ceramics, causes the appearance of the stresses in plates during their cooling, being stretchable ones at lower layers and constringent ones at upper layers. In fact, while moving away from plate substrate, it bends sideways, opposite to the substrate surface [6].

The presence of internal stresses is indicated in work [7], based on the experiments of thermal annealing of specimens at 1800C. The annealing has caused the redistribution of micro-hardness: it increased for 2.5 times for the deposition surface face, and decreased for 2.9 times from substrate face. Other macro-properties (the density, the TCLE, the heat conductivity) have not changed. Therefore the authors of [7] suppose that the micro-hardness is the measure of its micro-stresses in the carbon-glass ceramics. Consequently, the micro-hardness cannot completely reflect those sides of structure, which determine the properties of carbon-glass ceramics. At the same time, the thickness and surface plate distribution of this index are contributing itself to the heterogeneity of material.

For improving the material; properties the technology of obtaining the carbon-glass ceramics in reactor of 200 mm in internal diameter has been developed. He cylindrical substrate is made of GMZ graphite, densified with pyro-carbon. That has neared the substrate and carbon-glass ceramics TLCE. The creation of boundary layer has removed the adhesion. Thus the internal stresses were withdrawn. The productivity of technological process has increased; the price has decreased [8].

The average values of characteristics of USP carbon-glass ceramics, determined by accepted techniques, are given in table 1 for measurement at room temperature. They are compared to the properties of early obtained carbon-glass USB-15 ceramics at the same conditions, taken from work [9].

The obtained characteristics correspond to the indices of technical conditions for carbon-glass ceramics [10]. They meet requirements for the actuated elements of artificial heart valve.

As seen from table 1, the modification of technological process has resulted in obtaining more dense industrial material. Consequently, due to the presented bond [11], the USP carbon-glass ceramics has the higher strength, dynamic modulus of elasticity and thermal conductivity. The value of TCLE has not changed.

It should be noted that such a high value of dynamic modulus of elasticity was not obtained in work [12] even for the density of experimental USB-15 specimens USB-15 of 2.03 g/cm^3 of density.

For the pyrolytic isotropic graphite (PIG), obtained by the analogous technological scheme, but without boron, for the same density, the strength characteristics (the bending and pressure breaking point, the dynamic modulus of elasticity) are approximately for 2.5 times as lower [13].

The USP carbon-glass ceramics properties anisotropy was estimated as the ratio of measurements of micro-hardness for two perpendicular directions – for perpendicular and parallel direction to the substrate for the random samples, picked from different production lots. The obtained ratio was within 0.95-1.05, i.e. that confirms the isotropy of material. The same result was obtained while studying the specimens for the bending load. Hence the prismatic specimens of dimensions of 2×2×20 mm were cut in parallel and perpendicular (along the chord) directions to reactor axis. The obtained ratio is within 0.94-1.02.

Table 1: Properties of USB-15 and USP Carbon-Glass Ceramics

Characteristic	Index	
	USB-15	USP
Lattice parameter *c*, nm	0.6856	0.688-0.690
The sizes of crystallites, nm		
diameter L_a	15	9.0-10.0
height L_c	13	9.2-9.3
The degree of graphitization γ, rel.unit.	0	0
Density, g/cm^3		
bulk density, d_k	1.84	1.9-2.05
bottle density, d_b	1.95	2.0
X-ray density, d_x	2.215	2.216
The porosity, %		
mercury accessible porosity, Π	6.2	0-5
total porosity, Π_t	17	8-14
The breaking point, Mp		
the stretching breaking point, σ_s	87-123	—
the bending breaking point, σ_b	180-250	300-360
the pressure breaking point, σ_c	380-550	650-720
The dynamic modulus of elasticity *E*, Gp	19.5	22.9-23.1
Poisson's ratio μ, rel.unit.	—	0.27
The micro-hardness *H*, Gp	1-3	1-1.5
The specific electrical resistivity ρ, μΩ·m	12-18	14-15
The heat conductivity* λ, W/mK	14-17	23-25
TCLE* α, μK^{-1}	5.5-5.7	5.5
The mass loss rate** ω, 10^{-7} g/cm^2·s	1.2	0
The gas-permiability coefficient***, 10^{-6} cm^2/s	1.0	—

* TCLE is given for interval 20-2000C; the thermal conductivity – for room temperature.

** At oxidizing in the air at 600C.

*** For the air.

The indices shown in Table 1 are presented for room temperature of measurement (except the TCLE). With increasing the measurement temperature from the room temperature up to 1500-1700C, the indices increase (including the thermal conductivity as well) [6]:

the bending breaking point (Mp)	from 200 up to 310;
the pure bending sagging (mm)	from 0.8-1.1 up to 1.4-1.5;
the thermal coefficient of linear expansion (μK^{-1})	from 4.2 up to 5.5;
the heat conductivity (W/m·K)	from 15-20 up to 35.

At the same time, while heating the carbon-glass ceramics above 1800C, the intensive catalytic graphitization (due to the boron) results in rebuilding the turbo-strative structure, followed with sharp change of all properties. Thus, for example, the strength goes down for the extent [12].

In accordance to the accepted technology, at the production, every element of artificial heart valve is subjected to testing by means of the load with bending load up to 50% of the material breaking point. The aim of the test is to cull the products with the cracks. As determined, such a preliminary loads did not cause decreasing of this index at the beyond (after load) experiments. Thus obtained results coincided to the results of experiments for the specimens, which were preliminarily loaded to 356±29 and 345±24 Mp, respectively.

The carbon-glass ceramics valve element is the semicircular plate of variable section, having the lugs on the edges. The area of the change of the lug with to the flat is the concentrator of the stresses. There is a large probability of material failure first. In this connection, the effect of concentrator of stresses upon the bending breaking point has been evaluated. The incisions that are the concentrators of the stresses were made on the polished prismatic carbon-glass ceramics specimens of 2×2×20 mm of dimension, chosen from different production lots.

They were of the equal depth, differed with radius of rounding. The part of the specimens of the radius of 0.15 mm was polished in the incision area. The specimens with the concentrators of the stresses together with the same pilot uncut specimens were investigated for the bending load. The obtained average results are presented in Table 2.

As seen from the presented data, the larger radius of the rounding is the more preferable. For $R>1$ mm, the influence of concentrator of stresses is insignificant. The influence of the latter also decreases at polishing the material in the incision zone.

Table 2: The Dependence of Ratio of the Bending Breaking Points (*K*) of Specimens without Incision (The Check Specimens) and the Specimens with the Incision – The Concentrator of Stresses vs. Rounding Radius (*R*)

R, mm	0	0.05	0.10	0.15	0.15*	0.30	0.50	1.00
K*	4.32	3.37	3.08	2.42	2.13	1.85	1.84	1.21
ΔK**	—	0.23	0.15	0.26	0.16	0.29	0.27	0.10

* - the polishing in the incision area.

** - the averaging deviation.

HOMOGENEITY OF MATERIAL

The homogeneity is the very important parameter. For evaluating this parameter, the statistical analysis of the measured parameters was carried out. The Gaussian distribution is accepted in domestic works. The investigation of the sample of 30 specimens is enough for representing all the entire assembly. The variation of the properties is evaluated with the following parameters: the maximum (M_{max}), the minimum (M_{min}) and average (M) values of a given property; the mean-square deviation from the average (S); the variation coefficient or the changeability coefficient (v). The durability of the loaded elements is connected with the latter – the increase of coefficient (for the given margin of safety) decreases the level of durability, i.e. the ability of detail to work at no-destruction.

In foreign works, the homogeneity of graphite is evaluated using statistical Veibull's durability theory. The homogeneity of structure is characterized by m homogeneity parameter in Veibull's distribution, connected, according to [14], with durability variation coefficient.

The experimental results of statistical investigation of some properties of carbon-glass ceramics USP are given in table 3.

The measurements of specimens, chosen from different lots, are used. The additional measurements, performed to other sample USP specimens, taken also from the production lots, have shown the analogous results – the density, the bending strength, microhardness (and their variation coefficients) were equal to 1.91 g/cm^3 (3%), 34.6 Mp (8.1%), 1.05 Gp (10%), respectively.

In table 3, the comparison statistical USB-15 carbon-glass ceramics indexes are taken from [9].

The data demonstrate the large dynamical modules of elasticity and microhardness of USP carbon-glass ceramics in comparison to early produced materials.

Table 3: Parameters of Statistical Distribution of Properties of USB-15 and USP Carbon-Glass Ceramics

Material	Characteristic	Index				
		d_k,g/cm^3	ρ, μΩ·m	σ_b, Mp	E, Gp	H, Gp
USB-15	M_{min}	1.837	17.9	200	15.6	1.35
	M_{max}	1.950	19.8	250	20.7	1.83
	M	1.780	15.6	180	12.8	1.07
	S	0.028	1.0	—	1.9	0.29
	v, %	1.5	5.6	—	12.0	21.0
USP	M_{min}	1.890	14.5	345	25.2	1.14
	M_{max}	2.020	—	415	26.9	1.51
	M	1.820	—	308	23.4	1.02
	S	0.050	—	24	1.3	0.13
	v, %	2.7	—	7	5.2	11.8

THE CORRELATION BETWEEN MATERIAL PROPERTIES

The microproperties of carbon materials are known to be mainly caused by their crystalline and porous structure. For mathematical describing their dependence vs. open porosity, there is widely proposed (see, for instance [11]) and widely used the expression:

$$P=P_0\exp(a\Pi), \qquad (1)$$

where P, P_0 are the properties of material with the pores and of "nonporous" material, respectively; a is the parameter depending on distribution of pores vs. equivalent radius sizes; $\Pi=1-d_k/d_p$ is the open porosity; d_k, d_p is the bulk and bottle porosity.

The parameter a is the constant for a given class of carbon material and is experimentally determined for each property. The early-evaluated USB-15 carbon-glass ceramics breaking points have yielded the value of 5. For dynamical modulus of elasticity the parameter a is usually the same. The preexponential term P_0 is usually caused by crystalline structure and microstructure of given material. In a semi-logarithmic co-ordinates the expression (1) is defined by line with slope ratio of a.

If now the average values of modulus of dynamical elasticity and bending resistance of specimens PGI, USB-15 and USP put onto the diagrams in dependence on their average total porosity, then in the semi-logarithm co-ordinates, through the experimental points for the carbon-glass ceramics the lines of equal slope ratio of 4.5-5 can be drawn (Fig.1 and 2). The line corresponding to USP elasticity modulus is higher than for USB-15. The difference of these materials in parameter P_0 can be completely related to the difference in the microstructure achieved due to modification of technological process. For carbon-glass ceramics, as well as for

USP obtained nowadays, the parameter P_0 is higher. For bending resistance (Fig.2) such a difference, within dispersion, is not observed.

The point, representing the boron free PGI, is lower than experimental points for USB-15. The dotted line supposedly of the identical slope ratio is directed through it, since the considered materials belong to the same class. The presence of carbide phase B_4C in USB-15 is the reason of increasing its strength.

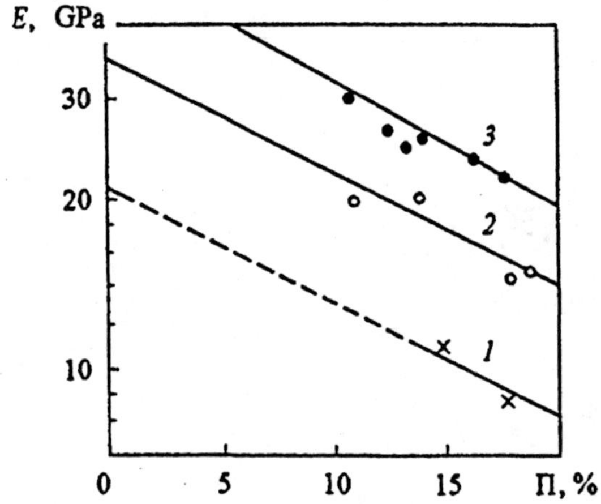

Figure 1. Dependence of dynamic modulus of elasticity vs. total porosity in materials: 1 – PGI [9], [13]; 2 – USB-15 [6]; [9] – USP.

The correlation in the strength properties for carbon-glass ceramics (the bending breaking point and the modulus of elasticity) with the total porosity is similar to the early determined relation for the constructional graphite of burnt coke-stuff electrode technology [11].

The statistical treatment of results of measurement for two samples of specimens from different USP lots has yielded the coefficients r of pair correlation between density and micro-hardness, being equal to 0.68 and 0.39. In second case the correlation is practically absent, probably, because of micro-stresses influencing. Therefore the additional statistical investigation of material is required.

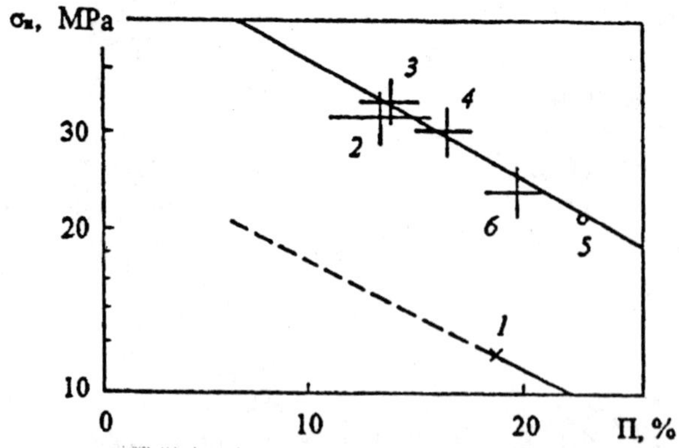

Figure 2. Dependence of the bending breaking point vs. total porosity in materials: 1 – PGI [13]; 2 – USP [8]; 3 – USP [16]; 4 – USB-15 [9]; 5 – USP in a given work.

From (10) it is followed that there must be a correlation between micro-properties as well. Really, the coefficient of pair correlation (*r*) between the pressure breaking point and micro-hardness appears to be 0.9. The correlation between bending breaking point and micro-hardness is exhibited in work [15]. However, the calculation of pair correlation between micro-hardness and bending breaking point, performed in a given work for choosing the industrial specimens, has shown that r=0.42, i.e. the correlation appears to be slight.

CONCLUSIONS

1. The modification of technology of obtaining the carbon-glass ceramics has led to increasing the strength and dynamic modulus of elasticity and has increased the homogeneity of material.
2. The correlation of bending breaking point and dynamic modulus of elasticity with the total porosity is similar to this for burnt coke-stuff graphite.
3. For using the micro-hardness as the criterion for estimating the operational performance of carbon-glass ceramics the additional investigation is required.

REFERENCES

1. Volkov G.M., Dobrova N.V., 3aharova E.N, i dr. Uglerodnyy material dlya iskusstvennogo klapana serdtsa. V sb. Uglerodnye konstruktsionnye materialy, vyp. 14, M.: Metallurgiya, 1979, s. 96-99.
2. Volkov G.M., Kalugin V.I., Syskov K.I. Dokl. AN SSSR, 1968, t. 183, № 2, s. 396-397.
3. Volkov G.M., Kurkin A.B., Leontev E.A. Himiya tverdogo topliva, 1970, № 3, s. 100-106.
4. Volkov. G.M., Kalugin V.I. Himiya tverdogo topliva, 1972, № 5, s. 90-93.
5. Agafonov A. U, Dobrova N.V., Fadeev A.A., Zavalishin N.N. J.of Advanced Mater., 1995, no. 2(l), p. 61-65.
6. Ostrovskiy B.C., Virgilev Yu.S., Kostikov V.I,, Shipkov N.N. Iskusstvennyy grafit. M.: Metallurgiya, 1986, 272 s.
7. Kurolenkin E.I., Virgilev Yu.S., Chugunova T.K., Muraveva E.V. Izv. AN SSSR, Ser.neorgan. mater., 1985, t. 21, № 2, s. 308-311.
8. Tatarinov V.F., 3olkin P.I. Ogneupory i keramicheskie materialy, 1999, № 1-2, s. 25.
9. Virgilev Yu.S. Neorganicheskie mater., 1995, t. 31, № 1, s. 3-18.
10. Utoesitall, TU-9493-001-27771122-97.
11. Virgilev Yu.S. Himiya tverdogo topliva, 1973, № 5, s. 102-105.
12. Virgilev Yu.S. Atomnaya energiya, 1984, t. 57, vyp. 5, s. 353-356.
13. Lebedev I.G. Razmernaya nestabilnost grafitov pri vysokotemperaturnom obluchenii v yadernom reaktore. Preprint, Dimitrovgrad, NIIAR-19 (822), 1991, 32 s.
14. Chernyavey A.N., Gusman I.O., Chernyh V.A. Tsvetnye metally, 1980, № 3, s. 59-62.
15. Kurolenkin E.I., Virgilev Yu.S., Volkov G.M. i dr. Issledovanie struktury i svoystv uglesitalla. V sb. Struktura i svoystva uglerodnyh materialov. M.: Metallurgiya, 1984, s. 115-119.
16. Tatarinov V.F., 3olkin P.I. Issledovanie svoystv uglesitalla. Ogneupory i keramicheskie materialy, 1999, № 3, s.17.

ABOUT THE AUTHORS

***Valeriy Fedorovich Tatarinov** – the SPP "Pyrouglerod" (Penza), a director. An expert in area of materials science.*

***Yuri Sergeevich Virgil'ev** – the SRI Grafit (Moscow). A Doctor of Technical Sciences, a professor. A chief of laboratory and chief research officer. An expert in area of materials science.*

***Sergei Vasilievich Evdokimov** – the Research Association "Medinzh" (Penz). Director General. An expert in area of designing the artificial heart valve.*

Calorimetry of Phase Composition of Copolymers: Butilmethacrylat and Methacrylic Acid – as Binder in Therapeutic Systems

V.F. Uriash, T.E. Knyazeva, V.A. Izvozchikova and N.Yu. Kokurina

The heat capacity has been measured and the differential thermal analysis has been carried out for the series of buitilmethacrylat and methacrylic acid copolymers of different composition, which can be used as the binder in therapeutic systems (TS). The multi phase copolymers have two or three glass-transition temperatures depending on the composition. The thermodynamic characteristics (an entropy, enthalpy and Gibbs function) of the studied materials and the isotherms of their thermodynamic functions are compared.

Introduction

Presently, the interest to the medical product of prolonged action, in which the polymeric carrier is able to deliver the drug into the sick organism by a given program. Because of strict requirement demanding to polymeric matrixes in therapeutic systems, the correct match for them is the actual problem. Recently we [1-3] have shown that butilmethacrylat (BMC) and methacrylic acid (MAA) can be used as the binder in TS. However, it was necessary to choose the optimal ratio of monomers that the obtained copolymer provides the necessary adhesion of TS to the substrate and skin as well as the required withdrawal time of TS medical product.

In this connection we measured the heat capacity in region of 80-330 K and the differential thermal analysis (DTA) has been carried out in the region of 80-500K for the series of copolymers BMC-MAA containing 7-37 mol.% of acid.

Materials and Technique

The BMC-MAA copolymers were obtained by means of deep conversion suspension polymerization. The monomers were cleaned by means of fractionating at the lowered pressure. As chromatographic analysis has shown, the amount of major component in the selected fractions of monomers was 99.9 mol.%. The refraction index n_D at 293 K for BMC and MAA was 1.4240 and 1.43114, respectively, that coincides with known literature data [4]. The benzoyl peroxide was used as the initiator. The copolymerization was carried out at 343-353 K during four and half hours. After that the copolymers were washed out with distilled water and dried. The composition was determined by titration of the MAA carboxyl groups.

The heat capacity (C^o_P) was measured in vacuum adiabatic calorimeter described in work [5]. The studied specimens, containing 7.6; 28.5; 37.3 mol.% of MAA, and mass of 2.8044; 3.3137; 2.4363 g, respectively, were placed in platinum ampoule of volume of 7 cm^3. The temperature was measured with resistive thermometer. The calorimeter equivalent energy was determined by measuring the heat capacity of calorimetric ampoule filled with helium up to pressure of 8.5 kp. It did not exceed 40% of C^o_p of filled calorimeter in the studying temperature range. For verifying the technique for absence of systematic bias we have determined $C^0_p=f(T)$ of etalon benzoic acid of K-1 specification. The obtained results coincided with literature data [6] with accuracy of 0.3% in region of 80-330 K. The temperature increases for 1.0-3.0K in experiments on measuring the heat capacity, therefore the amendment for curvature of function $C^o_p=f(T)$ was not introduced. The deviation of experimental data from the averaging curve was within 0.2%.

The DTA was carried out to the plant described in [7]. The experiment was carried out in helium atmosphere. The quartz was used as the etalon in experiment. The buttons of specimen and etalon are ~0.5 g. The heating rate in experiments is 5 K/min.

DISCUSSING THE RESULTS

The obtained results are presented in Fig. 1-3 and in Table.

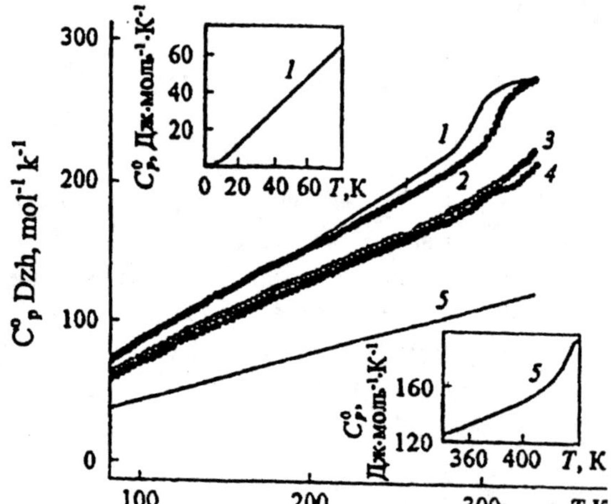

Figure 1. Temperature dependence of heat capacity of PBMS (1)[7], PMAA (5)[8] and series of BMS-MAA copolymers containing different MAA concentrations, mol.%; 2 – 7.6; 3 – 28.5; 4 – 37.3.

As seen in Fig. 1, the temperature dependence C^o_p of copolymer has the typical appearance for amorphous polymers. It monotonically increases right up to the beginning of transition of polymer from glassy state to high-elasticity state, and after, the anomalous increasing of C^o_p, connected with devitrification. As demonstrated by us [1-3], the copolymers BMS-MAA, containing acid of 20-37 mol.%, exhibit three devitrification temperature regions for average temperatures of T_{g1}, T_{g2} and T_{g3} (Fig. 2,3) in thermogram. The glass-transition temperatures of the homopolymer T_g (PBMS) and T_g (PMAA) were obtained by DTA technique and compared to literature data on heat capacity [8,9]. As seen form fig.3, they correlate to each other.

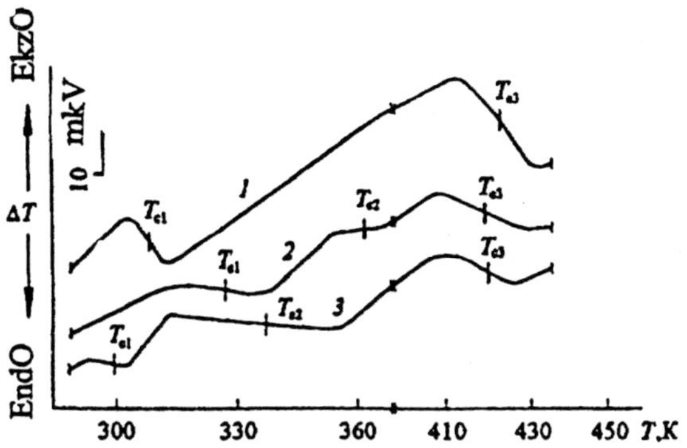

Figure 2. Thermograms of series of BMS-MAA copolymers containing different MAA concentration, mol.%: 1 – 7.6; 2 – 28.5; 3 – 37.3.

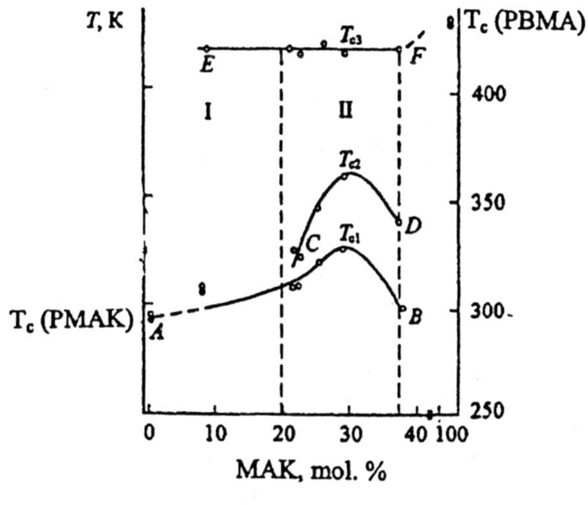

Figure 3. Concentration dependence of glass-transition temperature (T_{g1}, T_{g2}, T_{g3}) obtained from DTA(°) and heat capacity (•), of series of BMS-MAA copolymers. In diagram: I is the two-phase region; II is the three-phase region.

On the curve $C^o_p=f(T)$ of specimen of copolymer with MAA of 28.5 mol.% (Fig.1, curve 3) the beginning of first devitrification (T_{g1}) is observed, which does not finish at 330 K. The copolymer with MAA of 37.3 mol.% (Fig.1, curve 4) exhibits first devitrification (T_{g1}) and the beginning of second one (T_{g2}).

The specimen containing the acid of 7.6 mol.%, in comparison to other specimens in thermogram (fig.2, curve 1) exhibits two devitrification temperature regions (T_{g1} and T_{g3}). The first devitrification is also observed on the curve $C^o_p=f(T)$ (fig.1, curve 2).

The obtained results demonstrate the multiphase character of BMS-MAA copolymers (fig.3). As indicated in work [3], it can be connected with heterogeneity with respect to composition of copolymers BMS-MAA. The authors of work [10], which studied physic-chemical properties of methyl methacrylate copolymers (MMA) with butilcrylate and MMA with ethylhexylacrylate have come to same conclusion. The first and second devitrification temperature regions (T_{g1}, T_{g2}) relate, obviously, to the devitrification of BMS-MAA copolymers fortified with MAA to the less (curve *AB*) or the large degree (curve *CD*). The second phase appears only for concentration of MAA larger than 20 mol.%.

As seen in fig. 3, the curves AB and CD have the maximum at 28.5 mol.% of MAA. Such a character of dependence can be caused by two factors. The increase of number of MAA chains in copolymer causes, as a rule, the increase of T_g on one hand. On other hand, due to good solubility of acid in water, the further increase of concentration of MAA in primary blend favors to increasing its amount in water. As the consequence, the amount of MAA is redistributed in polymer fractions. The increase of specific amplitude of relaxation transformation T_{g3} and decrease of this value for T_{g1}, T_{g2} at the concentration of MAA larger than 28.5 mol.% is the confirmation of such interpretation. The 37.3 mol.% of MAA copolymer, created under these conditions, is the most heterogeneous one, as it consists of phases of practically two homopolymers PBMS (T_{g1}=299.5 K) and PMAA (T_{g3}=418 K) and copolymer BMS-MAA (T_{g2}=337 K).

The high-temperature copolymer transformation (T_{g3}) relates, obviously, to devitrification of phase of PMAA. The constancy of T_{g3} in the studied region indicates the incompatibility of phases. The fact that T_{g3} is for 15 K lower than T_g for homopolyner, is connected, possibly, with plasticizing effecting on the phases of the lower T_g or the water on the PMAA. However, it requires additional investigations.

For enriching the information on character of structure formation in copolymers BMS-MAA, the thermodynamic characteristics (the heat capacity, the entropy, the enthalpy and Gibbs function) of copolymers at constant temperatures and in dependence on composition (Table) were compared. For calculating the thermodynamic functions, the obtained dependencies $C^o_p=f(T)$ of copolymers were extrapolated to 0 K by Kelyl-Parks-Huffmann method [11]. For samples with MAA of 28.5 and 37.3 mol.%, the coefficients *A* and *B* in equation $C^o_p=C^o_{p,e}$ (A-B·T) were fitted thus that its decision coincided to experimentally determined value C^o_p in range 80-100 K with accuracy of ±0.35%. Hence, $C^o_{p,e}$ is the heat capacity of etalon substrate being allied to studied copolymers, measured from temperatures near 0 K. The PMAA was given as the etalon substance, the heat capacity of which is determined in work [8]. A=0.9239 and 0.6528; B=0.0005642 and 0.001289, for copolymers with MAA of 28.5 mol.%, respectively.

The heat capacity of 7.6 mol.% acid specimens below 80 K coincided with C^o_p of PMA within experiment accuracy. The thermodynamic functions $H^0(T)$-$H^0(0)$ and $S^0(T)$-$S^0(0)$ of BMS-MAA copolymer specimens were calculated by integrating the curves $C^0_p=f(T)$ and $C^0_p=f(\ln T)$, respectively, and Gibbs function $G^0(T)$-$H^0(0)$ was derived from Gibbs-Helmholtz equation.

The obtained concentration dependencies of isotherms C^o_p have shown that up to 200 K they are additive, then at 250 and 300 K the curves deflect from additive to negative direction and at higher temperatures the curves are reversed. Such a character of isotherms is explained by transformation of polymers from glassy state into high elastic in different temperature ranges. The enthalpy isotherms are practically additive ones in studied temperatures range, and the dependencies of entropy and Gibbs function have a small positive deflection from additive.

For the MMA-MAA copolymers the isotherms deflect in negative direction from additive dependencies [9]. Authors of [9] explain this by inter-chain interaction by means of hydrogen bonds in the acid monomer links. In studied BMS-MAA copolymers the polymer chain consists of BMS links, giving the elasticity for the molecule, and links of acid, the carboxyl groups of which are able to form the hydrogen bonds. Obviously, the observed positive deflection of isotherms of entropy from additive dependencies is connected with predomination of effect of elastic BMS links in copolymer macromolecules.

Table: The Heat Capacity and Thermodynamic Functions of BMS-MA Copolymers Containing the Acid of 7.6(I), 28.5 (II), 37.3 (III) Mol.%.

T, K	C^0_p, J/mole·K			$H^0(T)$-$H^0(0)$, kJ/mole			$S^0(T)$-$S^0(0)$, kJ/mole·K			-[$G^0(T)$-$H^0(0)$], kJ/mole		
	I	II	III	I	II	III	I	II	III	I	II	III
50	42.46	38.15	38.01	0.9183	0.8332	0.8330	29.07	26.45	26.44	0.5351	0.4892	0.4892
100	85.60	75.64	71.40	4.150	3.702	3.608	72.13	64.70	63.57	3.062	2.678	2.748
150	121.3	106.0	101.2	9.355	8.254	7.920	113.8	101.2	98.13	7.721	69.25	6.799
200	153.6	14.25	128.8	16.23	14.25	13.69	153.2	135.5	131.1	14.41	12.85	12.54
250	188.8	162.7	157.4	24.76	21.65	20.82	191.1	168.4	162.8	23.02	20.45	19.89
300	231.1	196.6	192.5	35.12	30.60	29.47	228.8	201.0	194.3	33.51	26.69	28.82
330	277.0	225.2	214.3	42.88	36.90	35.53	253.4	220.9	213.5	40.74	36.01	34.93

The ratio of those factors favors in creation of structure, providing the optimal properties, allowing to use it as a binder in TS (a good adhesion to the skin, necessary withdrawal time of medical product from TS).

CONCLUSION

The obtained results demonstrate the multiphase character of BMS-MAA copolymers, which, according to our view, can be caused by their composition heterogeneity.

The copolymers of composition of two-phase region I in diagram (fig.3) is the most appropriative as the binder in TS, as they provide good skin adhesion of TS and the required withdrawal time. The appearance of third phase (fig.3, region II) reduces the elasticity of copolymer and thus worsens physic-mechanical characteristics of binder.

Sincere appreciation is expressed to professor Sevast'yanov V.I. et al for investigation of adhesion and diffusion properties of binder.

REFERENCES

1. Uryash V.F., Izvozchikova V.A., Knyazeva T.E., Sevastyanov V.I., Iosilevich I. N. Mezhdunarodnaya konferentsiya po fundamentalnym problemam nauki o polimerah (k 90-letiyu akad. V.A. Kargina): Tez.dokl. M.: Rossiya. 1997, s. S3/88.
2. Ur'yash V.F., Izvozchikova V.A., Knyazeva T.Ye., Myasnikova LB., Losilevich I.N. Int.Conf. on Calorimetry, Experimental Thermodynamics, Thermal Analysis: Abstracts. Poland. Zakopane: 1997, p. 172.
3. Uryash V.F., Knyazeva T.E., Izvozchikova V.A., Maslova V.A., Iosilevich I.N. Zh.konversii vuzovskoy nauki: issledovaniya i razrabotki v oblasti medtehniki i biotehnologii, 1997, № 10, s. 74-75.
4. Lipatov Yu.S., Nesterova A.E., Grischenko T.M., Veselovskiy R.A. Spravochnik po himii polimerov. Kiev: Naukova dumka, 1971, 543 s.
5. Rabinovich I. B., Sheyman M. S., Nistratov V. P., Kamelova G. P., Зorin A.D. Zh. fiz. himii, 1985, t. 59, s. 2414-2417.
6. Rybkin N.G, Orlova M.P., Baranyuk A.K., Nurullaev N.G., Rozhnovskaya L. I. Izmerit, tehn, 1974, № 7, s. 29-32.
7. Uryash V.F., Mochalov A.N., Pokrovskiy V.A. Termodinamika organich. soedin.:Mezhvuz.sb. Gor-kovskiy gos.un-t, 1978, vyp. 7, s. 88-92.
8. Lebedev B.V., Kulagina TT., Smirnova N.N., Kiparisova E.G., Novoselova N.V. Zh. fiz. himii, 1995, t. 69, s. 581-591.
9. Pavlinov L.I., Rabinovich I.B., Okladnov N.A., Arzhakov S-A.Vysokomolek. soedin., 1967, t. A9, s. 483-487.
10. Ovchinnikov E.Yu., Gorelov Yu.P. Vysokomolek. soedin., 1988, t. B3O, s. 677-679.
11. Parke G., Haffman G. Svobodnye energii organicheskih soedineniy. M.: GIHL, 1936, 214 s.

ABOUT THE AUTHORS

Vladimir Fayvishevich Uriash *– NII of Chemistry of Nizhegorod State University (Nizhniy Novgorod). A candidate of Chemistry, a research officer. An expert in area of physic-chemistry and thermodynamics of physiologically active substances and polymers.*

Tatyana Evgenieva Knyazeva *– NII of Chemistry of Nizhegorod State University (Nizhniy Novgorod). A candidate of Chemistry, a research officer. An expert in area of synthesis of high-molecular compounds and investigation f their physic-chemistry properties.*

Valentina Alekseevna Izvozchikova *– NII of Chemistry of Nizhegorod State University (Nizhniy Novgorod). A candidate of Chemistry, a research officer. An expert in area of structure of physic-chemistry properties of (co)polymers and their compositions.*

Nadezhda Yurievna Kokurina *– NII of Chemistry of Nizhegorod State University (Nizhniy Novgorod). A junior research fellow. An expert in area of physic-chemistry and thermodynamics of physiologically active substances and polymers.*

Igor' Natanovich Iosilevich *– NII of Chemistry of Nizhegorod State University (Nizhniy Novgorod). A post-graduate student. An expert in area of physic-chemistry and thermodynamics of physiologically active substances and polymers.*

DISCRETE DEPENDENCE OF POWDER STEELS PROPERTIES ON POROSITY

V.N. Antsiferov, S.N. Bobrova, S.A. Oglezneva and Yu.I. Ragozin

The influence of porosity parameters on mechanical properties of powder steels, obtained by traditional and high-energy technologies, has been studied, and the anomalous (discrete) character of their dependence has been determined. The phonon failure theory, based on assuming the quantum character of levels of power-consuming failure, is proposed to be one of the explanation of this effect. As demonstrated, the use of high-energy mixing (mechanical alloying) of low-alloyed powder steels, providing high level of mechanical properties of constructional materials, is the promising method.

INTRODUCTION

The powder metallurgy holds one of the key positions among promising resources-economy technologies. The scientific direction of powder metallurgy is based on the disciplines as: physic-chemistry and mechanics of deformable disperse bodies, the solid state physics, theoretical and applied materials science. From this point of view the powder metallurgy has the wide possibilities for forecasting the physic-mechanical properties of new materials, providing with required operational performances of units of different purpose.

The pores of powder body are the important hereditary constituents of structure, significantly influencing on mechanisms of failure of constructional materials. On one hand, the decrease of effective cross-section of a loaded material can cause the brittle failure, on other hand, the blunting and the breaking of the created cracks by pores provides a significant reliability of products [1,2]. The dependencies of failure on porosity are ambiguous: according to [3], the strength and plastic properties of powder materials are monotonically increasing with the porosity. In contrast to that, there are the data about appearance of brittle failure while approaching the nonporous state of powder objects (4-10% of pores), because of the change of stressed state near pores [4,5] and redistribution of dopants on the grain boundaries in the closed porosity structure, as well as the debatable evidences about nonmonotonic crack resistance in other intervals of porosity (about 15%) [6].

At any case, the porosity most significantly influences on failure viscosity of powder steels [7]. The determination of the nature of the phenomenon will allow forecasting the conditions of appearance of anomalies that is of doubtless practical interest. Presently, the correlation between porosity and K_{1c} index (failure viscosity) in a given works were considered only from positions of general significance of porosity, while the mechanism of material resistance to external loads are obviously determined also by the dispersity, the pore shape and the distance between them.

EXPERIMENT

We have studied the influence of porous structure on crack-resistance and other mechanical properties of powder alloys of two types (table): the ordinary powder steel Pa-ZhGrD3 (the mixing of the initial powders has been carried out in the shifted axis mixer for 8 h), and "the mechanically alloyed" steel PK10M, produced using high-energy powder fragmentation method in X38H00 planetary-motion mixer for 25 min. From thus produced mixtures the billets were pressed by cold method in metallic molds at pressure from 200 up to 1000 Mp, and were sintered in atmosphere of dried hydrogen for the dew point of $t_{d.p.}$=-30C at temperature o5f 1200C during 2 and 3 h. The cooling rate was 50 degree/min. The choice of steels is dictated by good properties, by the simplicity and the developed technology for a known and widely used steel PA-ZhGrD3, as well as by the high physic-mechanical characteristics and advantageous steel structure, obtained by mechanical alloying method [8,9] for PK10M steel.

Table: The Chemical and Granulometric Compositions of Investigated Materials

The steel	Amount of steel elements, mass.%				Particle size, μm
	C	Cu	Mo	Fe	
PA-ZhGrD3	0.9	3.0	–	96.1	40-200
PK10M	0.1	–	1	98.9	6-10

The notice. The initial materials are:

- the PZhR 2.200.28, PZhR 3.200.28 iron, GOST 9849-86;
- the S-1 graphite, OST 6-09-431-75;
- the mixed carbon-free ferromolybdenum, TU 4759-70E, ST SEV 495-87;
- the PMS-1 copper, GOST 4960-75.

The bulk porosity was determined by hydrostatic method (GOST 18898-73) for every specimen. The characteristics of porous structure of powder materials were determined by metallographic method for magnification of 100-200 to the microscope "NEOPHOT-32" with the system of image digital processing SIAMS: the area was chosen on the thin-section field containing from 200 up to 1000 pores, those porosity coincides to the accuracy of 0.5% with porosity, determined by hydrostatic method. In the chosen area, the following parameters were determined:

- the pore size is the average diameter, measured for 20 directions (from 0 up to 171^0);
- the factor of pore shape,

$$F_c = \frac{2\sqrt{\pi S}}{P}$$

where S is the pore area, P is the pore perimeter, F_c=1 for the circle;

- the specific surface is the ratio of total perimeter of pores and their total area;
- the fractal dimension (the parameter characterizing the structure fragmentariness) was determined by enlargement analysis net method;
- the average effective interparticle distance was measured to two perpendicular directions and was chosen for the most probable one.

The mechanical characteristics ($\sigma_{0.2}$, σ_v, δ, K_{1c}) were determined in the stretching experiments (GOST-18227-85) and three-point bending (GOST 25.506-85) in 3-5 specimens.

THE RESULTS OF INVESTIGATION

The microstructure of PA-ZhGrD3 steel after sintering is the pearlite+thin torn cement net, for the average diameter of grain of d_{av}=40 μm; the PK10M steel, alloyed with ferrite, d_{av}=9 μm.

Because of different size of initial particles of powder mixtures, and because of difference in processes of interparticle jointing of two investigated materials, their porous structure are significantly different, fig. 1. Thus, for the PK10M steel (fig.1*c*), the visual distribution of pores is more homogenous than for PA-ZhGrD3 steel (Fig. 1*a*); the shape is spherical (F_c=0.8), the pore size distribution histograms, obtained for a given parts, cover the wide range of diameters for PA-ZhGrD3 (Fig.1*b*) and the narrow one for PK10M (Fig.1 *d*). The shape of pores F_c for specimens PA-ZhGrD3 steel little depends on porosity, being 0.7-0.8.

The crack trajectory in metals has the fractal nature [10], and the porosity plays important role in its distribution, therefore the investigation of dependence of fractal porosity index on its total value is of great interest. As seen in Fig. 2*a*, for both steels, it increases with porosity, thus for the equal Π_{tot}, the fractal index is larger for mechanically alloyed material. It is easy to explain the experimental data, issuing from the conception of fractals as the rare and branchy objects [1]. The fractal dimension is the quantative measure of branchy structure, which is defined by expression:

$$\rho(R) \sim \frac{M}{R^d} \sim R^{d_f - d}$$

where ρ, *R* and *M* is the density, radius and object mass, respectively, *d* is the space dimension, d_f is fractal dimension. In our case the fractal index characterizes the porous structure. For the identical porosity, it is larger for alloy steels, (fig. 2*a*) that is confirmed by results of determining the specific pore surface, fig. 2*b*.

With decreasing the bulk pore amount, caused by increasing the pressure, in the process of evolution of porous structure at the sintering (the absorption of small pores with large one, the healing, the creeping and the pores going out, etc [12]) there is observed the monotonic increase of average distance between pores in both steels, thus in the mechanically alloyed steel the distance is less due to large number of disperse pores for identical values of total porosity.

The analysis of dependence of mechanical properties of powder materials on porosity exhibits the non-monotonic change of breaking point, breaking viscosity and percent elongation of powder steels in wide porosity interval, thus all the properties are changing symbasily and the minimums for the identical values of porosity on the specimens for different experiments are observed, Fig. 3. The unusual (in comparison to dense materials) phenomenon of simultaneous growth of parameter K_{1c} and the strength is explained by the dominating significance of factor of resistance to the spreading of the moving instable crack [7,13].

The properties of the mechanically alloyed PK10M steels, despite of significantly less amount of the carbon, is for the extent larger than properties of ordinary powder PA-ZhGrD3 steel.

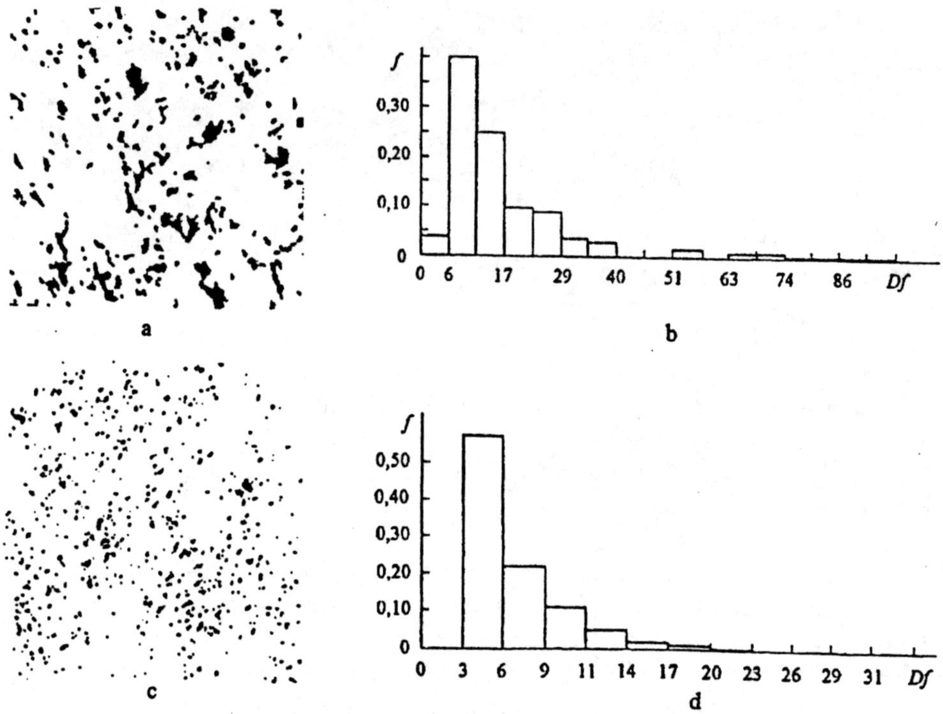

Figure 1. The porous structure (a, c) and sizes of pores (b, d) of sintered steels PA-ZhGrD3 (a,b) and PK10M (c, d).

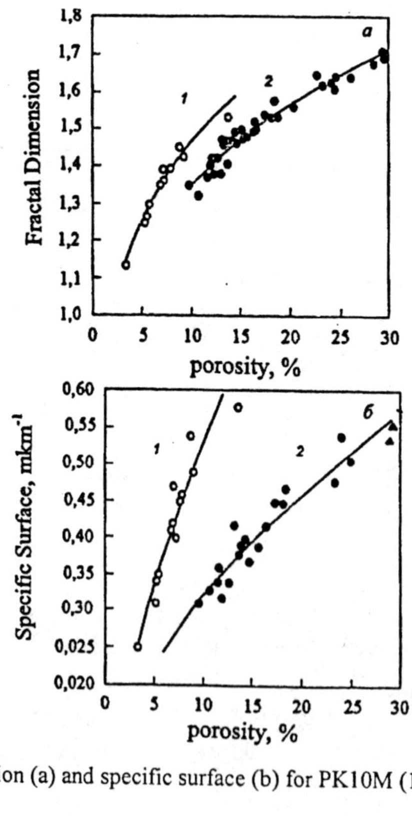

Figure 2. The fractal dimension (a) and specific surface (b) for PK10M (1) and PA-ZhGrD3 steels (2).

As the different characteristics of porous structure were obtained by varying the pressure, it is of great interest to reveal the correlation between physic-mechanical characteristics of the steels and the value of cold densification, during which the hereditary prerequisite of the quality of the jointing of the contacting surfaces of powder particles occur, i.e. the material strength properties. It has been determined that these dependencies have non-monotonic character, fig. 4. The violation of monotony of the change of breaking viscosity and the breaking point on the pressure for both steels are observed at 6-6.5 t/cm^2 and 8 t/cm^2 (by the way, the mostly spread and recommended industrial pressures for the iron powders).

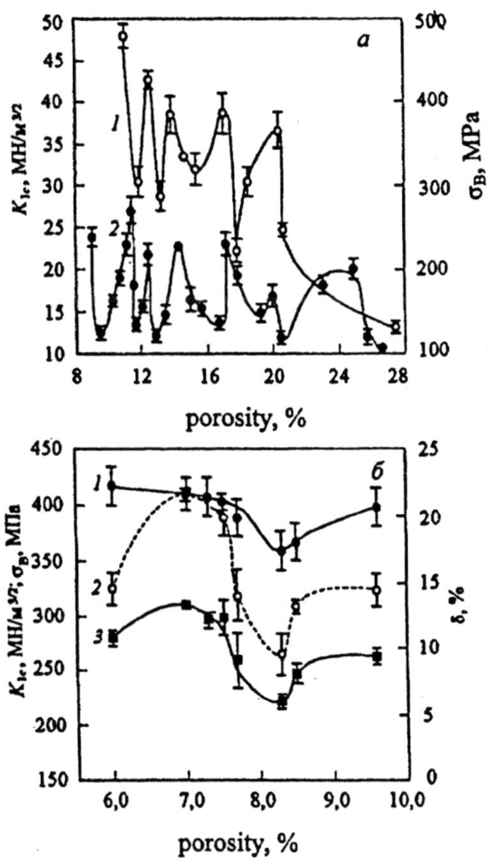

Figure 3. The dependencies of breaking point (1), the failure viscosity (2), the percent elongation (3) vs. the porosity of PA-ZhGrD3 (a) and PK10M (b) steels.

DISCUSSING RESULTS

The powder materials, obtained from initial mixture by the pressing and beyond high-temperature treatment (sintering), are the two-phase materials containing the pores and the dense phase. That determines their complex behavior under external loads and excepts the unambiguity interpretation of the dependencies between their physic-mechanical properties and structural parameters.

Not excepting the influence of the structure factors, the porosity and the distance between pores appeared to be the most significant ones, the authors of present investigation propose to use the concept of phonon failure theory as one of the explanation of the non-monotony of dependence of mechanical properties [14, 15]. From this position, the power-consuming of alloy sharply decreases at a certain critical distances between defects for a given material due to appearance of internal additional forces. The critical distances between defects for alloys on the base of iron are: 37, 47, 74, 76, 81, 94, 107, 118, 152, 162, 177, 214, 236, 354 μm. If the proposed model is correct then two its important sequences, followed from it, are also correct.

First of all, there should be several "dips" of crack resistance, and secondly, the mechanical properties of powder materials must also non-monotonically depend on porosity, as the criterion K_{1c} according to the phonon failure concept [15], is connected with mechanical characteristics of equation [16]:

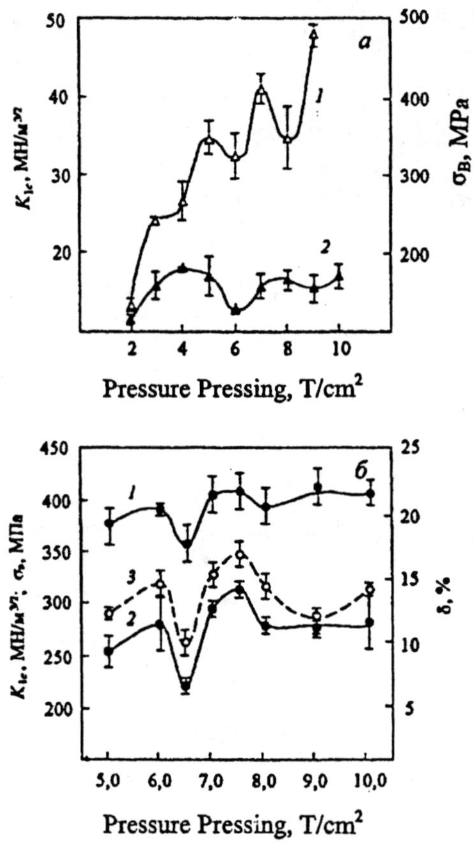

Figure 4. The dependencies of breaking point (1), the failure viscosity (2), the percent elongation (3) vs. the pressure of PA-ZhGr3 (a) and PK10M (b) steels.

$$K_{1c} = \sqrt{\frac{0.4 \bullet 10^{-3}\left[2(\delta_e E - \sigma_{0.2})(\sigma_{0.2} + 2\sigma_B) + 3\sigma_{0.2}^2\right]}{3(1-\mu^2)}}$$

where δ_e is a full (including elastic component) uniform deformation; $\sigma_{0.2}$ is the yield point; σ_b is breaking point; E is elasticity modulus; μ is the Poisson's ratio.

As in the powder materials the cracks expand in the dense component, it is logically to assume that basis parameter responsible for the "dip" of the crack resistance is the distance between pores. At the same time, taking into consideration the specificity of powder metallurgy technology when consolidation of material, pressed from the different particles, is reflected in the total volume of microvoids and their morphology, one should take into consideration the phenomenological aspect of this problem. Based on the above, the non-monotonic character of change of mechanical properties on porous structure indices of the investigated steels is caused by integral influence of energy and structure factors.

Thus, the obtained dependence of K_{1c} vs. the distance between pores, when it exceeds 70-80 μm (the porosity is less than 12-14%), confirms the phonon discontinuity failure theory – the "dips" of properties correspond to the critical values of microcrack length divisible by half-length of electromagnetic radiation in iron, fig. 5a (a lesser number of "dips" in the PK10M dependence is caused by lack of experimental data for corresponding values of porosity, Fig. 5b). The "dips" on the curve $K_{1c}=f(L)$ in the region of the distances between pores less than 60 μm (the porosity is larger 25%) are explained, obviously, basically by other reason: the low strength of interparticle contacts and by the unfavorable, the incise shape of some large pores.

In that, the mechanically alloyed PK10M steel has the advantage, caused by the uniform disperse porosity, due to which the failure power-consuming decreases (however, one can not ignore the positive influence of molybdenum), being not worse to K_{1c} for compound alloyed dense constructional materials.

Concerning the non-monotony of change of mechanical properties of powder steels with pressure, the coincidence of peak pressures in PA-ZhGrD and PK10M is the evidence that the "dips" and "peaks" of mechanical characteristics are not connected in this case with neither the obtaining mixture method nor granulometric composition, nor the quality of alloying element, but are determined by the nature of the material basis and can be connected with the quantum character of failure levels. On other hand, the recrystalization processes, developed at sintering the deformed jointing zones of the billets at their heating up to the sintering temperature followed with isothermal exposure, play the important role. It is known [17] that relaxation processes rate, the size of new deformed grains and the level of properties achieved after recrystallization essentially depends on the degree of preliminary plastic deformation.

In our case it is very difficult to separate the influence of different factors (the characteristics of pore space and kinetics of interparticle recrystallization), being changed with pressure of initial billets, on the processes of consolidation of powder billets at sintering, but the obtained experimental material confirms once again the inadmissibility of simplified extrapolation at obtaining the non-monotonous dependencies of properties of metallic alloys on both the technological parameters and the external loads to the beyond exploration.

CONCLUSIONS

1. The mechanical properties of powder steels depend non-monotonically on characteristics of porous structure that is caused by total influence of energy and morphology on the failure.
2. The dispersion of mechanical properties of powder products, which is often assumed as the properties instability of either material, may be of not random character but regular one. The determination of the similar dependencies for every concrete material will allow forecasting

the mechanical characteristics, and the strict observance of technological regimes – to avoid the known faulty properties.

3. The mechanical alloying method is the promising direction for obtaining the powder steels having favorable porous structure, providing the high failure power-consuming.

The work has been performed under support of grant RFFI #96-26-5.3-77: "The development of the scientific principles for improving the structure and heightening the failure power-consuming of economy-alloyed powders of alloys on the basis of iron".

A.S.Volkov and V.B. Kul'mentieva have taken part in the experiment.

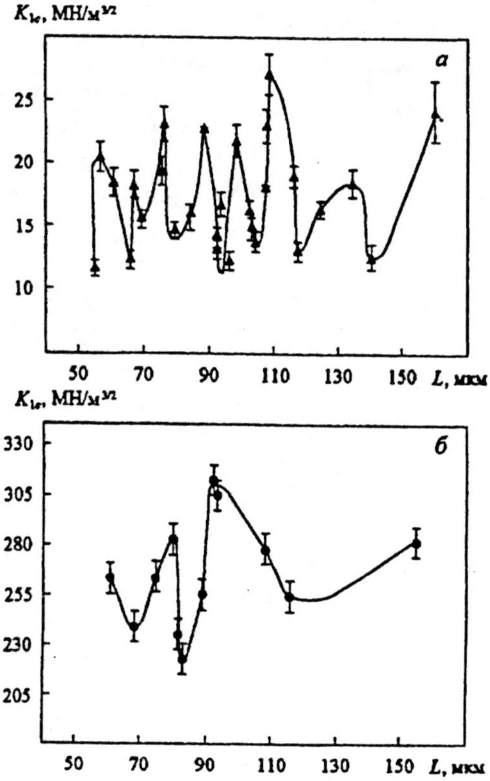

Figure 5. The dependence of failure viscosity of powder PA-ZhGr3 (a) and PK10M (b) vs. distance between pores of K_{1c}.

REFERENCES

1. Milman Yu.V., Leskovekiy A.M., Ivaschenko R.K. i dr. Poroshkovaya metallurgiya, 1994, № 12, s. 77-85.
2. Martirosyan N.S. Poroshkovaya metallurgiya, 1985, № 10, s. 97-103.
3. Andrievskiy R.A. Poristye metallo-keramicheskie materialy. M.: Metallurgiya, 1964, 187 s.
4. Drachinskiy A.S., Kuschevskiy A.E., Perepelkin A.V. i dr. Poroshkovaya metallurgiya, 1982, № 12, s. 80-84.
5. Firstov S.A., Podrezov Yu.N., Zherdin A.G. i dr. Poroshkovaya metallurgiya, 1988, № 1, s. 69-73.

6. Milman Yu.V., Ivaschenko R.K., Moskalenko N.P. i dr. Hladnolomkost poroshkovogo zheleza. Tez. dokl. XV nauchn.-tehn. konf. Poroshkovaya metallurgiya. Kiev: In-t probl. materialoved. AN USSR, 1985, s. 295-296.
7. Bernshteyn M.L., Kutser M.Ya., Platova S.N., Svoyatytskiy V.B. Vyazkost razrusheniya spechennyh materialov. Sb. Poroshkovaya metallurgiya, vyp. 1, Minsk, "Vysheyshaya shkola", 1977, s. 100-107.
8. Bobrova S.N., Mazein S.A. Otrabotka rezhimov tsementatsii poroshkovyh nizkomolibdenovyh splavov. Problemy razvitiya metallurgii Urala na rubezhe XXI veka: Sb. nauchn. trudov po materialam Mezhgosudarstvennoy nauchn.-tshn. konf. Magnitogorsk: RIO MGMA im. G.I. Nosova, 1996, s. 31-36.
9. Antsiferov V.N., Oglezneva S.A., Pescherenko S.N. Fizika metallov i metallovedenie, 1998, № 2, s. 98-104.
10. Emelyanov A.A., Portaov L.E., Pyshmintsev I.Yu., Pospelov Yu.V. Izvestiya vuzov. Chernaya metallurgiya, 1994, № 6, s. 43-47.
11. Guld X., Tobochnik Ya. Kompyuternoe modelirovanie v fizike. Chast vtoraya. M.: Mir, 1990, 391 s.
12. Geguzin Ya.E. Fizika spekaniya. M: Nauka, 1984, 312 s.
13. Antsiferov V.N., Shubin V.N., Cherepanova T.G. Izvestiya vuzov. Chernaya metallurgiya. 1977, s. 143-177.
14. Ragozin Yu.I. New fundamental fracture criterial. Proc. llth Eur. Conf. Fract., 1996, v. l, p. 289.294.
15. Ragozin Yu.I. Metally, 1996, № 6, s. 69.
16. Ragozin Yu.I., Antonov Yu.Ya. Problemy prochnosti, 1984, № 2, s. 28.
17. Gorelik S.S. Rekristallizatsiya metallov i splavov. M.: Metallurgiya, 1978, 568 s.

ABOUT THE AUTHORS

Vladimir Nikitovich Antsiferov *– The Republican Engineering-Technical Center for Powder Metallurgy and NII of Problems of Powder Technology and Coatings (RITC PM) at Perm State Technical University (PSTU). A professor, the Corresponding Member of RAS, a director. An expert in area of powder metallurgy, physics and chemistry of solid state.*

Svetlana Nikolaevna Bobrova *– The Republican Engineering-Technical Center for Powder Metallurgy and NII of Problems of Powder Technology and Coatings (RITC PM) at Perm State Technical University (PSTU). A Candidate of Technical Sciences. A research officer. An expert in area of physical metallurgy and powder metallurgy.*

Svetlana Arcadievna Oglezneva *– The Republican Engineering-Technical Center for Powder Metallurgy and NII of problems of Powder Technology and Coatings (RITC PM) at Perm State Technical University (PSTU). A Candidate of Technical sciences. A research officer. An expert in area of physical metallurgy and powder metallurgy.*

Yuri Ivanovich Rogozin *– The Perm State Technical University (PSTU), a Candidate of Technical Science. A professor. An expert in area of physics and mechanics of solid state failure.*

EFFECT OF $FeAl_3$ AND CARBON DOPING ON TRIBO-TECHNIC PROPERTIES OF SINTERED ALUMINUM BRONZE

E.N. Korosteleva, A.P. Savitskii and N.M. Rusin

The possibility of increasing the tribo-technic properties of the sintered aluminum bronze has been investigated. As shown, the additional doping by intermetallic particles positively influences on friction constant and the wear resistance of sintered material that substantiates using it as the anti-friction material in mid-loaded friction units.

INTRODUCTION

The most products of anti-friction destination, based on an iron and copper (first of all the sleeve bearings) are produced by powder metallurgy method. The ferrographite alloys and tin bronze belong to these materials. The relatively chip sintered alloys based on aluminum bronze, demonstrating a high operational performances at friction experiment, can hold a fitting place between these products.

The number of publications, devoted to investigation on the properties of alloys on the basis of aluminum bronze, is not large. Basically, there were investigated the high-dense alloys of complex composition, obtained by sintering under pressure, or undergone to densification [1-4]. The two-component aluminum bronze, sintered at temperature of tin bronze sintering, has the low mechanic properties [5]. In general, the properties of the sintered aluminum bronze, depending on the composition, were studied not well enough for speculating the optimal composition ensuring its high mechanic and tribo-mechanic properties.

The additional introduction of the solid lubrication particles and solid inert particles like the carbides, borides etc, increasing the wear resistance, are the known method of increasing the properties of anti-friction materials [6]. However, these dopants, as rule, resist to shrinkage at sintering, and for achieving the high mechanic properties such an alloys are subjected to additional heat densification that increases their costs [7,8].

According to literature [1,2,7,8], the intermetallic particles and the graphite are increasing the tribo-technic properties of alloys. Therefore the aim of present work is to investigate the effect of doping intermetallic $FeAl_3$ and graphite on the tribo-technic properties of sintered aluminum bronze. The choice of iron aluminide as the anti-friction dopant into sintered aluminum bronze is caused by that the precast BrAZh9-4 aluminum bronze of antifriction destination, used in the industry, contains from 2 up to 4 mass.% iron, which is insoluble in copper and forms solid intermetallic inclusions with the aluminum. The powder of this intermetallide is easily produced, as the $FeAl_3$ billets, sintered from the mixture of the most accessible and non-expensive metals – iron and aluminum, are spontaneously dispersed. The $FeAl_3$ particles are strongly bond with plastic matrix, being the solid solution of copper and aluminum, are insoluble in copper at sintering and, forming the solid inclusions of second phase, provides the high anti-friction properties of material.

THE MATERIALS AND INVESTIGATION TECHNIQUE

The alloys were obtained in vacuum furnace by sintering the pressed mixtures of the elemental PMS-1 copper and PA-4 aluminum powders. The $FeAL_3$ intermetallide and EUT-2 graphite powders were the dopants. The experiment on the friction was carried out to the SMT-1 friction machine for the scheme "disk-shoe", where the steel hardened disk of 45 steel was as the rider. The friction was performed for two slip velocity – 0.52 and 1.83 m/s over cooled lubricant. The load over the specimen was gradually increasing till the moment increases sharply and the gross wear occurs, or till the specimens breaks. All the specimens were undergone to the wear-in at the load of 100 N till stabilizing the friction moment. With aim of the comparison, the experiments were also carried out for the mold BrOTz-10-2 tin bronze.

THE RESULTS AND DISCUSSION

The dependencies of friction coefficient and temperature in the rubbing pair contact zone at slip velocity of 0.52 and 1.83 m/s vs. the load over sintered Cu-Al alloy and molded BrOZ-10-2 tin bronze are presented in Fig.1. The effect of graphite introduced into alloy on friction coefficient and load-carrying ability of sintered Cu-Al-C alloy is exhibited in fig. 2. The results of investigation of effect of particles of iron aluminide $FeAl_3$ on friction coefficient and load-carrying ability of aluminum bronze are presented in fig. 3. The curves of tribo-technic properties of sintered aluminum bronze vs. load over specimen at mutual introducing the articles of intermetallide and graphite into alloy are presented in fig. 4.

From comparison of experimental results on friction of sintered Cu-Al aluminum bronze and molded BrOZ-10-2 bronze it is seen that sintered alloy has a lower friction coefficient, however the maximally permissible load over it is less than for molded alloy. Thus, the increase of slip velocity decreases the load-carrying ability of sintered bronze. The decrease of permissible load on the specimen can be the sequence of a lower mechanical properties of sintered alloys in comparison to the molded one, because of presence of pores, being the elements worsening the mechanical properties of material. In process of friction, the role of pores is ambiguous. On one hand, they are the effective reservoir for liquid lubricant, which protects the friction surfaces from direct contact. On other hand, the pores as the stress concentrators are the places of the most probable crack nucleation. Their appearance and accumulation causes the premature failure of friction surface and separation of its fragments from basis material in the form of particles. Since in a given experiment the lubrication is present permanently on the friction surface, the efficiency of pores as lubrication sources appears to be low, whereas their weakening effect on the material remains.

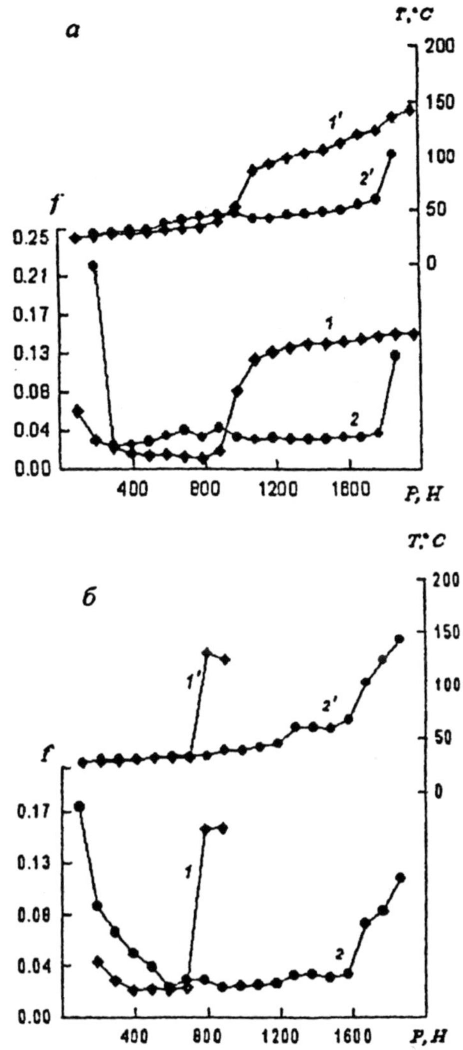

Figure 1. Dependence of friction coefficient (1,2) and temperature in contact zone (1',2') at slip velocity of 0.52 (*a*) and 1.83 m/s (*b*) vs. load over specimen: 1, 1' – the sintered Cu-Al alloy; 2, 2' – the molded BrOZ-10-2 tin bronze.

Other reason of a lower load-carrying ability of sintered Cu-Al alloy can be the absence in its composition of soft component, which would be as the lubrication when achieving the critical loads on friction surfaces. Thus in the separate places the oil film can be destroyed and the surfaces of riders are in dry contact. It is known that in such contacting place surfaces at friction the surface temperature significantly increases [9]. In the case of tin bronze in such situation its low-melting-point component overflows onto friction surface, forming the lubrication film, protecting the riders from direct contact, being as the lubricant [10]. At using the aluminum bronze such effect is absent.

From comparison of behavior of alloys Cu-Al and Cu-Al-C (fig.2) it is seen that graphite favors some increasing the load-carrying ability of sintered aluminum bronze at the friction. Thus the friction coefficient decreases a little.

Other method of increasing the load-carrying ability and wear resistance of friction materials is the introducing of inert solid particles [7,8]. In the case of good bond with matrix, such particles effectively strengthen the material. Besides, for the sufficient their number, they overflow under surface and resist to the direct contact. The liquid lubrication can easily circulate between such particles and bring the products out the friction zone, effectively reducing the temperature [7,8].

In present work, the iron aluminide particles were as the dopant in the sintered aluminum bronze. As $FeAl_3$ is the saturated with respect to aluminum, one can expect that introducing of these particles into alloy practically does not cause the change of matrix composition. One can assume that some amount of copper is dissolved in a given intermetallide nevertheless, providing strong bond of its particles with matrix. The results of experiment on friction of sintered alloys with intermetallide particles are presented in fig. 3. It is seen that introduce of alluminides favorably effect on friction coefficient. Its value in the wide range of loads are the lower than in the rest investigated alloys.

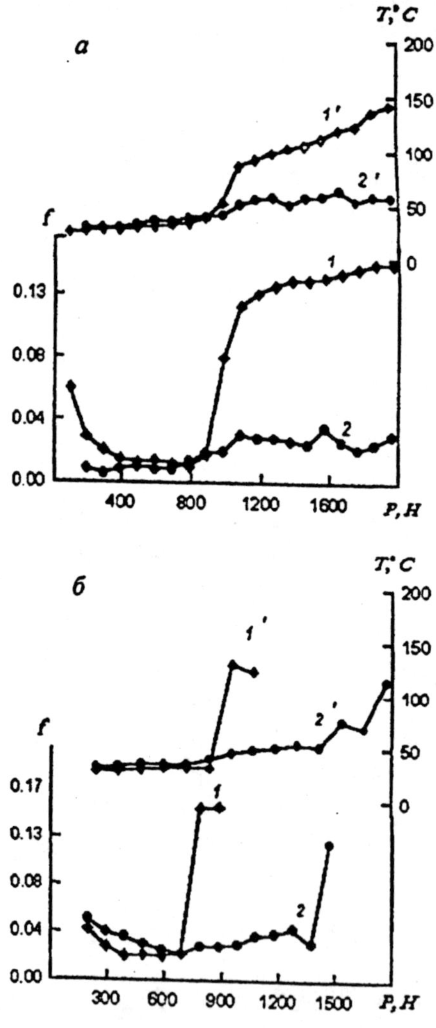

Figure 2. Dependence of friction coefficient (1,2) and temperature in contact zone (1',2') at slip velocity of 0.52 (*a*) and 1.83 m/s (*b*) vs. load over specimen: 1, 1' – the sintered Cu-Al alloy; 2, 2' – the sintered Cu-Al-C alloy.

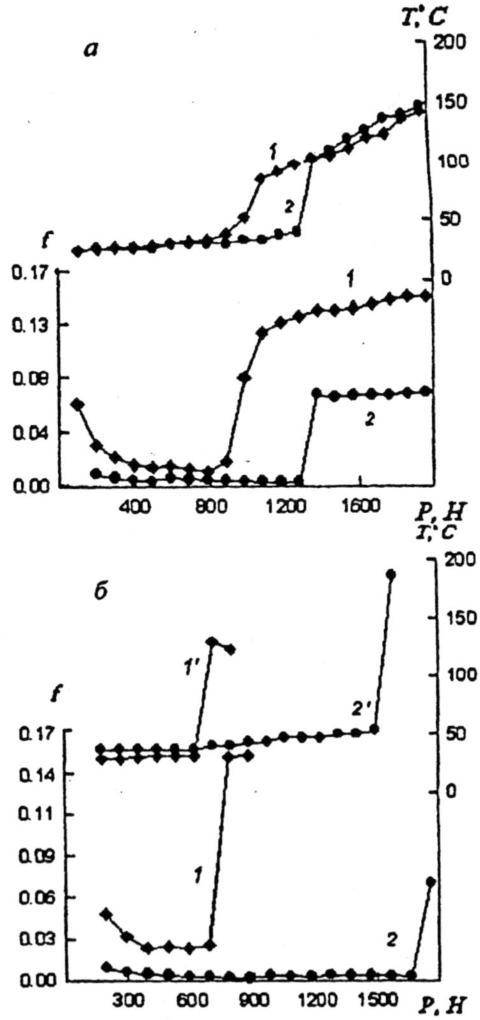

Figure 3. Dependence of friction coefficient (1,2) and temperature in contact zone (1',2') at slip velocity of 0.52 (*a*) and 1.83 m/s (*b*) vs. load over specimen: 1, 1' – the sintered Cu-Al alloy; 2, 2' – the sintered Cu-Al-$FeAl_3$ alloy.

As it is expected, the load-carrying ability of Cu-Al alloys with intermetallide inclusions under friction significantly increased. Moreover, at the slip velocity of 1.83 m/s as for the load-carrying ability the aluminum bronze contained the intermetallide (fig.3b, curve 2), exceeds the molded tin bronze in a given experimental conditions (fig. 1*b*, curve 2).

Obviously, the difference in the behavior of sintered aluminum bronze and the bronze strengthened with intermetallide $FeAl_3$ particles under friction is caused by that the mechanical properties of alloys with particles are the higher than for the alloy without them. For example, the hardness of strengthened material is 1210 Mp, whereas the hardness of Cu-Al double alloy is not more than 700 Mp. The pressure breaking point in material with intermetallic particles increased for 900 Mp in comparison to matrix material. The bronze plasticity, containing the intermetallides, thus decreased, but this decrease does not cause essentially the load-carrying ability of material.

Based on the obtained results one could expect that joint introduction of intermetallide and graphite, as well as the separate introduce, is effecting positively on the tribo-technic properties of sintered aluminum bronze. The results of joint effect of $FeAl_3$ particles and graphite on the friction coefficient and temperature in friction zone are presented in fig. 4. Regardless of expected result, the tribotechnic characteristics of alloy, containing jointly graphite and intermetallide, essentially decreased. There can be several reasons of increasing the friction coefficient and decreasing the load-carrying ability in Cu-Al-C-$FeAl_3$ alloy.

First, the positive influence of graphite as the solid lubrication on the load-carrying ability of bronze can be neglected by the particles of graphite in the sintered alloy, likely to the pores, effect negatively on the mechanical properties [11, 12]. As the aluminum bronze is the system for which the liquid phase disappears at sintering [13], it is hardly expect the good adhesion on the "graphite-matrix" boundary. The imperfect interface, being the place of nucleation of cracks, decreases the mechanical properties of the bronze. One should not except the possibility of formation of brittle aluminum carbides on the surface of graphite particles due to its interaction with aluminum. The presence of carbides must also decrease the alloy durability [14].

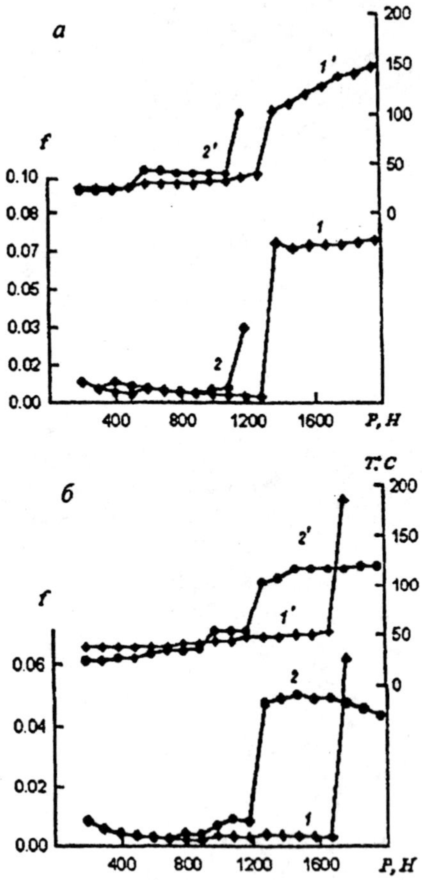

Figure 4. Dependence of friction coefficient (1,2) and temperature in contact zone (1',2') at slip velocity of 0.52 (*a*) and 1.83 m/s (*b*) vs. load over specimen: 1, 1' – the sintered Cu-Al-$FeAl_3$; 2, 2' – the sintered Cu-Al-$FeAl_3$-C alloy.

Secondly, the particles of graphite and intermetallides at the mixing and beyond formation can create the contacting but not bound between themselves the inclusion of second phase, which resist to the shrinkage of alloy at sintering [15]. Besides, the presence of the chain of such particles in the sintered material favors spreading the fatigue cracks created under friction. As the result, the larger regions are invoked into the chafe process, which create the larger chafe particles and cause the sooner chafe of experimental material.

CONCLUSIONS

From results, presented in a given work, one can conclude, that sintered aluminum bronze can effectively substitute a more expensive tin bronze under low and tempered load friction. Thus, in contrast to tin bronze, the introduce of graphite into sintered Cu-Al alloy is not efficient under conditions of presence of constant lubrication at low loads. The most promising material for tribo-technics is the sintered aluminum bronze with dopants of internmetallic $FeAl_3$ particles.

REFERENCES

1. Kryachek V. M., Shvedkov E. L. Poroshkovaya metallurgiya, 1969, № 9, s. 60-65.
2. Kryachek V. M., Shvedkov E. L. Poroshkovaya metallurgiya, 1969, № 10,s. 66-70.
3. Antsiferov V. N., Shatsov A. A. Trenie i iznos, 1996, t. 17, № 2, s. 213-217.
4. Perelman O.M., Bobrova S.N., Melnikov M.Yu. i dr. Tsvetnye metally, 1990, № 7, s. 94-97.
5. A.s. 1346334 (SSSR). Sposob izgotovleniya spechennyh izdeliy iz alyuminievoy bronzy. Savitskiy A.P., Emelyanova M.A. Opubl. v B.I., 1987, № 39.
6. Shvedkov E.L. Poroshkovaya metallurgiya, 1983, № 6, s. 37-51.
7. Karapetyan G.H., Akopov N.L. i dr. Poroshkovaya metallurgiya, 1987, № 4, s. 75-79.
8. Karapetyan G.H., Akopov N.L. i dr. Poroshkovaya metallurgiya, 1987, № 5, s. 85-88.
9. Albegachiev A.Yu., Braun E.D., Sorokin G.M. Raschet i ispytanie friktsionnyh par. M.: Mashinostroenie, 1974, s. 102-107.
10. Зozulya V. D. Ekspluatatsionnye svoystva poroshkovyh podshipnikov. Kiev: Naukova dumka, 1989, 288 s.
11. Karapetyan G.H., Akopov N.L. i dr. Poroshkovaya metallurgiya, 1988, № 6, s. 85-88.
12. JhaA. K., Prasad S. V., Upadhyaya G. S. Powder Metal., 1989, U. 32, № 4, R. 309-312.
13. Hansen M., Anderko K. Struktury dvoynyh splavov. M.: Metallurgizdat, 1962, t. 1, 608 s.
14. Gorbylevskiy M. A. Problemy mashinostroeniya i avtomatizatsii, 1989, vyp. 26, s. 45-52.
15. Rusin N. M., Savitskiy A. P. Poroshkovaya metallurgiya, 1994, № 3-4, s. 42-46.

ABOUT THE AUTHORS

Elena Nikolaevna Korosteleva *– the Institute of Physics of Strength and Materials Science of Siberian Branch of RAS (Tomsk). A research fellow. An expert in area of powder metallurgy.*

Arnold Petrovich Savitskii *– the Institute of Physics of Strength and Materials Science of Siberian Branch of RAS (Tomsk). A professor, a Doctor in Physics and Mathematics. A laboratory Chief. An expert in area of powder metallurgy.*

Nikolay Martemianovich Rusin *– the Institute of Physics of Strength and Materials Science of Siberian Branch of RAS (Tomsk). A Candidate of Technical science. A research officer. An expert in area of powder metallurgy.*

HEAT CONDUCTIVITY OF Cu_2-O-Cu CERMETS

V.V. Ivanov, V.I. Aletshenko and V.V. Slabko

The heat conductivity λ of specimens for all the stages of production of Cu_2O-Cu cermets, containing up to 40 mass.% Cu, obtained by means of deoxidizing the copper oxide, has been measured. The influence of composition and technological parameters, as well as the porosity of billets and finished specimen have been studied. The value λ and the heat conductivity matrix component progressively increase with C_{Cu} due to formation of "endless cluster" consisting of metallic particles. The dependence of heat conductivity and relative density of cermet have been defined: $\lambda=\lambda^0\varepsilon^{3.8}$.

The heat conductivity of the brittle compositions based on oxide ceramics which often work at high temperatures, significantly defines the important characteristic – the resistance to the heat impact. In particular, the Cu_2O-Cu cermets are the interesting materials in creating the inert high-temperature electrolytic anodes, thus the heat conductivity is the important physical parameter. On the other hand, the determination of heat conductivity is the physic-chemical analysis method giving the information on the structure of composition material, the phase connectivity, the porosity and their change in experiment.

In a present work, we have measured the heat conductivity λ of specimens for all the Cu_2O-Cu cermet production stages, obtained at deoxidizing the basis Cu_2O oxide, described in patent [1]. The measurement was carried out for temperatures of 20±2C by stationary method with the heat conductivity-measuring device ITEM-1M. The specimens of 8-10 mm in diameter and 7-10 mm in length were used. Every measurement was duplicated to the two-three specimens.

The "green" billets were densified at pressure of 3 t/cm^2 by two-sided pressing in steel mold, prepared from the mixture of copper oxide powder, deoxidizer (phenol-formaldehyde tar) and binder (polyvinyl-butyral). The powders, given in the necessary proportion, were subjected to wet treatment in planet-centrifugal mill at 20g (g is the acceleration of gravity). After drying, the prepared mixture was as the large aggregates from the small (2-5 μm) oxide particles glued with organic dopants. The relative density of specimens $\varepsilon=d/d_t$ was calculated with respect to theoretical density d_t of porousless material, calculated as the additive value of density of all the components.

The relative density of "green" billets was about 0.73-0.75, and their heat conductivity in studied range of binder amount and deoxidizer of 1.9-3.45 mass.% does not practically depend on amount of organic dopants and was within 0.69-0.84 W/(m·K) at average λ_{green}=0.74 W/(m·K).

It is interesting to study the behavior of value λ_{therm} of specimens after heat-treatment of billets at relatively low temperature, leading to the finishing of main processes of thermolysis of organic dopants and deoxidizing of oxide, but is not sufficient for consolidation of deoxidized copper films and the sintering of material. Such a heat treatment has been carried out by means of slow heating up to temperature of 700C followed with cooling. As appeared, the specimens of lower copper amount (~14 mass.%) did not change the sizes; starting from C_{Cu}=18 mass.%, the significant and the increased shrinkage occurred. Thus, the mentioned temperature is not sufficient for increasing the mobility of atoms in the oxide lattice and its sintering, but provides the necessary concentration of defects and diffusion in metallic phase. The shrinkage of specimens under heat treatment is that their porosity appears to be an identical for all cermets compositions, estimated as θ=40±5%. The porosity of the specimens after organic burning naturally increases due to burning of the dopants, as well as due to difference of mole volumes of copper and oxide at deoxidizing.

The measured λ_{therm} is presented in Fig.1 (curve 1). The heat conductivity of specimens after organic layer elimination is in the dependence on copper amount and significantly increases, but is relatively low 1.5-8.0 W/(m·K).

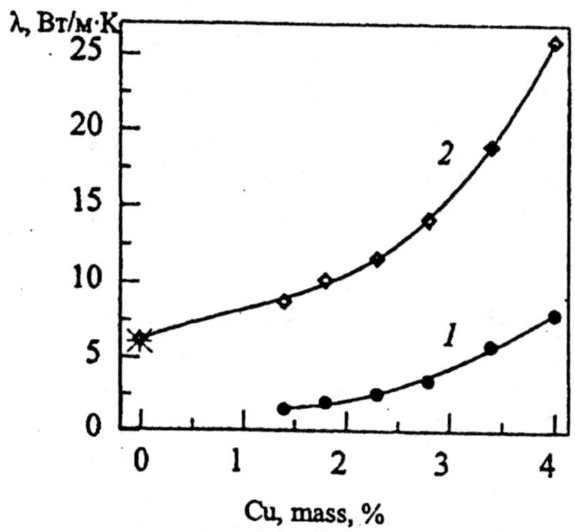

Figure 1. The dependence of conductivity of the annealed (1) and sintered (2) specimens vs. copper amount (λ=6.2 W/(m·K) at C_{Cu}=0 – according to [2]).

The sintering performed at temperature of 1323±10 K causes the increase of the heat conductivity increasing progressively with amount of metallic component. The experimental values λ of specimens sintered for 10 h are exhibited in fig. 1 (curve 2). The heat conductivity of all the compositions has increased for 3-5 times as much in comparison to the annealed material. At sintering, the value λ depends on technological regimes – the pressure P of specimens and the sintering duration τ (fig. 2). Only for small amount of copper, with increasing the P and τ the heat conductivity of specimens increases in whole parameter range. The larger amount of metal in material, the more sharply the increase of λ is altered with its decrease. The analysis of the whole experimental data has defined that the phenomenon is connected with the peculiarities of the considered structure and is determined by increase of diffusion porosity at high-temperature heat-treatment.

The presence of a great number of specimens, having the different porosity for equal phase contents, had allowed finding the correlation dependencies, binding the heat conductivity and the relative density of material. Due to its practical importance, the bind λ-ε (or λ-θ) of powder material always attracts the attention of researchers. There are the great number of experimental and theoretical ratios, describing the influence of porosity on the properties of the type of one-phase bodies. The most known equation is M.Yu.Balshin's one:

$\lambda=\lambda^0\varepsilon^2$ and $\lambda=\lambda^0\varepsilon^{3/2}$, (1)

where λ^0 is the coefficient of heat conductivity of porousless material of a given composition. The equation is valid, respectively, for matrix (the communicating pores) and static (isolated pores) systems. They usually well describe the experimental data [3]. However, concerning the compound bodies consisting of the particles of sharply different conductivity, the given correlations are exhibited to be not suitable. In order to describe the experimental data in such cases, the analogous equations with higher exponents are usually used.

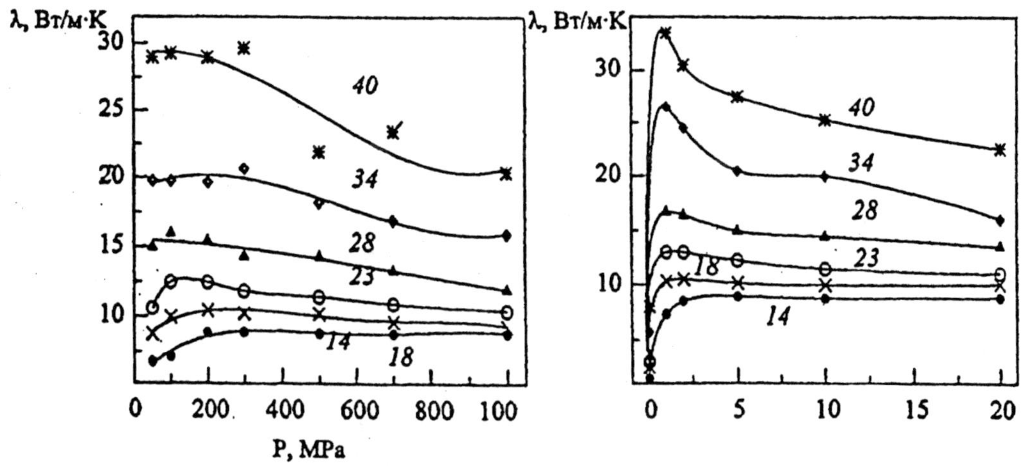

Figure 2. The effect of technological factors (pressure and sintering time) on the heat conductivity of sintered specimens (the digits on the curves are the amount of copper in cermet, mass.%).

In fig. 3 and in table the measurement results binding λ and ε are summarized

$\lambda=\lambda^0\varepsilon^n$ or $\lg\lambda^0+n\lg\varepsilon$, (2)

separately for every cermet composition. The δ correlation coefficients of corresponding curves are given as well. The clear correlation λ-ε is observed, thus the exponent for whole amount of copper is significantly higher than for theoretical one. The available data exhibit the independence or slight dependence of exponent n on the cermet composition. The average value is n=3.8.

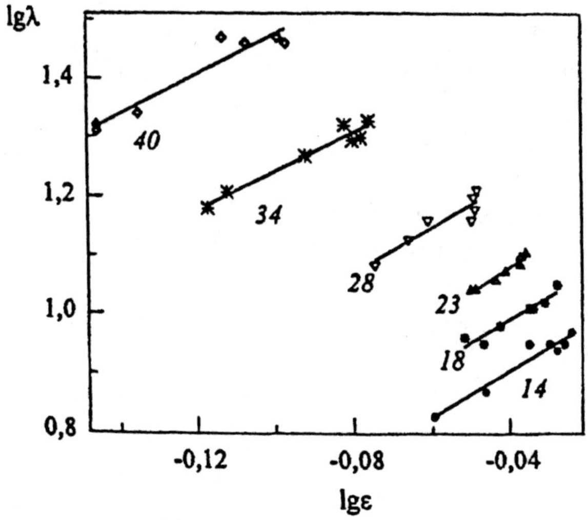

Figure 3. The correlation of heat conductivity and relative density of cermets.

Table: The Heat Conductivity of the Annealed Porousless Billets and the Constants of Correlation Λ-E

Amount, Cu mass%	λ^0_{therm}, W/(m·K)	λ^0, W/(m·K)	n	Δ
14	10.6	11.5	3.88	0.94
18	14.3	14.1	3.88	0.92
23	18.5	17.8	4.36	0.94
28	24.7	24.0	3.84	0.86
34	41.1	37.2	3.27	0.95
40	55.9	66.1	3.42	0.92

The values of coefficient λ^0 of hypothetic porousless material, obtained from fig. 3 and corresponding to experimental equations (2), are presented in fig. 4 as the curve 1, depending on copper bulk composition. The estimated values of effective heat conductivity of cermet due to theoretical equations based on generalized conductivity [4,5] are presented in the same diagram. For the compound with closed inclusions (the statistic system, isolated inclusions), the Odolevskiy's equation is used, (curve 2):

$$\frac{\lambda_{stat}}{\lambda_1} = 1 - x_2\left(\frac{1}{1-\nu} - \frac{1-x_2}{3}\right)^{-1} \qquad (3)$$

where $\nu=\lambda_2/\lambda_1$, x_2 is the bulk portion of copper, indices 1 and 2 correspond, respectively, to oxide and metal:

$$\frac{\lambda_{matrt}}{\lambda_1} = C^2 + \nu(1-C)^2 + \frac{2\nu C(1-C)}{\nu C + 1 - C}$$

where 0 C 1, C=0.5+$A\cos\varphi/3$, 270^0 $\varphi \leq 360^0$ and in our case, when 0 x_2 0.5, A=-1, φ=arcos(1-$2x_2$).

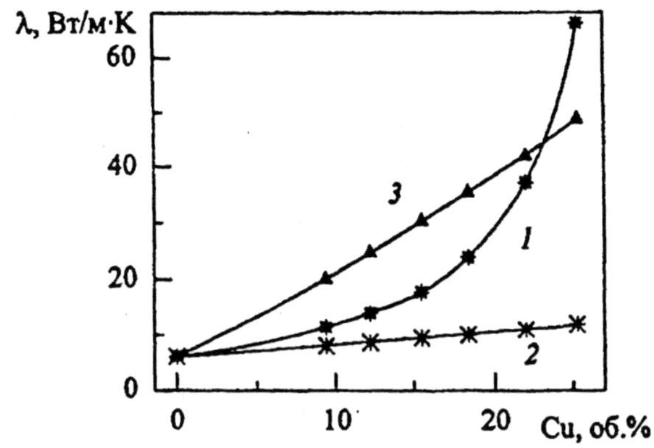

Figure 4. The heat conductivity of cermets (ig.3), reduced to the porousless state (1), the calculation for static (2) and matrix (3) systems.

It should be noted that calculated curves 2 and 3 (fig.4) look like the linear dependencies in graph. It is true at the beginning and in not wide interval x_2. As seen from figure, the curve λ-C_{Cu} of porousless cermet passes basically between curves corresponding to two "mechanisms" of heat conductivity: "statistic" and "matrix". Thus, for low amount of copper, the λ values are near the "statistic" dependence and increase progressively after C_{Cu}~15 vol.%. Issuing from the leaking theory concept [6] for the bulk portion of well conducting component x_2=0.15±0.03, the isolated clusters (IC) consisting of particles of this component are amalgamating into the endless cluster (EC), forming the bound structure and the effective conductivity of the system start increasing. According to such concept, with increasing the amount of copper in Cu_2O-Cu cermet, the sizes and the branching of the closed inclusions of IC increase, the contribution of matrix component of heat conductivity increases as well. For x_2>0.15 this portion is the determinative. The larger ratio $\nu=\lambda_2/\lambda_1$, the more sharp transformation IC EC. For example, the measurement of ω conductivity of studied cermets has shown that for ratio $\nu=\omega_2/\omega_1$ 10^9 (in contrast to $\nu=\lambda_2/\lambda_1$ 60) the sharp jump for 5-6 degrees of value in narrow range Δx_2 near values of x_2=0.16±0.01 is observed in dependence ω-C_{Cu}.

Let us discuss the above value λ_{green}=0.74 W/(m·K), which is for the degree lower than the heat conductivity of copper oxide $\lambda(Cu_2O)$=6.2 W/(m·K) [2]. Taking into consideration that in three phase oxide-organics-pores system the portion of organic component changes within 0.067-0.089, and its heat conductivity is 0.25 W/(m·K) [7], the apparent deflection of experimental values from the data estimated by the additive rule is observed. The calculation of λ_{green} (by Balshin's equation (1) for material with connecting pores) to the porousless state yields λ^0_{green}=2.1 W/(m·K). In such a two-phase body, the bulk portion of organic component increases up to 0.115-0.154, but the heat conductivity is disproportionately low. The organic dopant in a given case is distributed as the system of adsorbed continuous polymer films on the oxide particles isolating these particles each of other. Therefore the contribution of dispersed oxide phase significantly decreases by thermal resistance of polymer films. Such effects are well known and are exhibited in crowded polymer and glue compounds [7,8]. The porousless powder body, consisting of Cu_2O particles isolated each other by polymer film continuous phase can be considered as the system with the close inclusions (the structure like emulsion) that allows using the equation (3) for estimating the λ^0_{green} value. The calculation yields λ^0_{green}=2.5 W/(m·K) that is near the above estimation from equation (1).

On other hand, the validity of the made estimations and corresponding equations can be raised by inverse calculation of λ^0_{green} coefficient of porousless specimen. For θ=40%, the pores form the connecting system that allows using the equation (4) with taking into consideration the heat conductivity of porousless two-phase matrix λ^0_{green}=2.5 W/(m·K) and inclusions – the pores fulfilled with air (λ=0.025 W/(m·K)). The calculation yields: λ_{green}=0.67-0.8 W/(m·K). Thus, the calculated coefficients are in the interval of experimental values. That is the evidence of closeness of the results from equations in the case when the structure of powder body corresponds to the models, described by these equations. In the considered above example (fig.4) such a good coincidence of experiment and calculation is not observed because the cermet structures changing due to the increase of copper amount.

The close value of porousless of all heat-treated specimens allows calculating the λ^0_{therm} of porousless material of corresponding composition using the defined correlation $\lambda=\lambda^0\varepsilon^{3.8}$. The calculated values for θ=40% are presented in table. As seen, they slightly differ from those obtained for sintered specimens. This is the evidence of both the correctness of the assumptions about the completeness of thermolysis process of organic dopants and the deoxidizing, and the validity of the obtained correlation λ-ε. Besides, the coincidence of λ^0_{therm} and λ^0 demonstrated the slight dependence of heat conductivity on the morphology and the structure of dispersed components of cermet: if in the annealed specimen the copper is in the form of thin films or slightly consolidated fine-disperse inclusions then, in finished material, is as the large enough particles. It is well seen in fig. 5 which has been obtained by means of the computer processing of the considered cermet structure pictures of Cu_2O-34Cu compound. The copper inclusions are marked with black. The slight dependence of λ vs. the structure of isolated inclusions is not surprising – it has been defined and postulated in the theory of generalized conductivity [5,6].

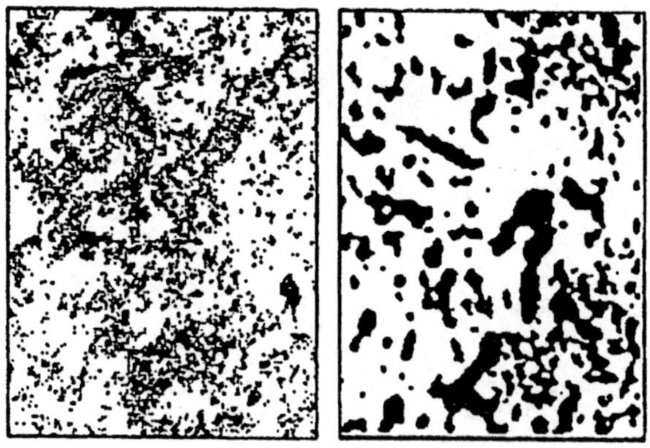

a б

Figure 5. The structure of copper inclusions in the annealed (a) and sintered (b) specimens. The magnification ×200.

In conclusion, one should underline that the coefficient of heat conductivity of studied cerments is relatively small, essentially increases with copper amount and with decreasing the material porosity. The comparison to the heat conductivity of heat-resistant constructional ceramics based on the Al_2O_3 and MgO having the approximately identical value λ≈15-40

W/(m·K) allows concluding that the resistance to the thermal stresses of cermets Cu_2O-Cu is determined by, first of all, physic-mechanical properties which do not yield in the presented ceramics from the considered point of view. In fact, the quality experiments have demonstrated high enough heat resistance of this compound: the anode specimens of dimension of 10×20×70 mm stood the direct introduction of the electrolyte into the melt (800-850C) without primary heating.

REFERENCES

1. Polozhitelnoe reshenie po zayavke № 96106857/02, prioritet ot 4.04.96. Sposob izgotovleniya nesgoraemyh anodov. Ivanov V.V., Ivanov Vl.Vl., Polyakov P.V., Blinov V.A., Kirko V.I., Savinov V.I..
2. Bogoroditskiy N.P., Pasynkov V.V. Materialy radioelektronnoy tehniki. M.: Vysshaya shkola, 1969, 422 s.
3. Andrievskiy R.A. Poroshkovoe materialovedenie. M.: Metallurgiya, 1991, 203 s.
4. Dulnev G.N., Зarichnyak Yu.P. Teploprovodnost smesey i kompozitsionnyh materialov. L.: Energiya, 1974, 264 s.
5. Dulnev G.N., Novikov V.V. Protsessy perenosa v neodnorodnyh sredah. L.; Energoatomizdag, 1991, 247 s.
6. Shklovskiy B.I., Efros A.L. Uspehi fiz. nauk, 1975, t. 117, № 3, s. 401-434.
7. Popov V.M. Teploobmen cherez soedineniya na kleyah. M.: Energiya, 1974, 304 s.
8. Chalyy V.T, Ositinskaya T.D., Ryabinina O.B. Teploprovodnost almazopolimernyh kompozitov. V kn. Protsessy vzaimodeystviya na granitse razdela faz. Kiev: ISM AN USSR, 1982, s. 109-116.

ABOUT THE AUTHORS

Viktor Vladimirovich Ivanov *– the Research Physic-Technical Institute of Krasnoyarsk State University. The associate professor, a Candidate in Chemistry. An expert in area of powder materials science.*

Vadim Ivanovich Aletshenko *– the Research Physic-Technical Institute of Krasnoyarsk State University. The research fellow. An expert in area of solid state physics.*

Vitaliy Vasilyevich Slabko *– the Research Physic-Technical Institute of Krasnoyarsk State University. The Doctor in Physics and Mathematics. An expert in area of surface physics.*

PROPERTIES OF TEFLON COMPOSITE MATERIALS OBTAINED BY EXPLOSION PRESSURE

N.A. Adamenko, Yu.T. Trykov and E.V. Sedov

The aim of this work is to study the influence of explosion treatment on the structure and physic-mechanical properties of polymers and polymer-based composites, taking into account composition, initial porosity of compound blend and filler dispersity. The poly-tetra-fluorine-ethylene powders (teflon-4) with metallic (the bronze, the copper and the iron) and non-metallic (aluminum oxide) fillers of dispersity of 14-250 μm of concentration of 3 up to 50% have been investigated.

The strength and the stability of the adhesion bonds in the polymer-filler system are the main factors that determine the working capacity of polymer matrix filled composite materials [1-3]. In order to increase these parameters, the technological processes of obtaining the composite materials (CM) and activating the polymer matrix by means of the chemical modification [1], the mixing of polymers [2], the use of surface-active materials [3], the radiation and the ultra-violet and the other effects [4] are used. As determined by [5-7], the explosion treatment of polymers and powder composite materials improves their physic-mechanical properties due to the surface activation of particles, caused by friction, by the intensive plastic deformation, by the heating and the ionizing of the air in the pores.

The aim of a given work is to study the effect of shock-wave treatment (SWT) on the structure and physic-mechanical properties of polymers and polymer-based composite materials with taking into consideration the composition, initial porosity of composite blend and the filler dispersity.

The poly-tetra-fluorine-ethylene (teflon-4) powders and its mixtures with metallic (bronze, copper, iron, nickel) and non-metallic (aluminum oxide) fillers of dispersity of 14-250 μm of concentration from 3 up to 50% have been studied.

After dry mixing of the polymer powders and the filler, the composite blends were subjected to shock-wave pressure by the flat load scheme of pulse duration of 40 μs and maximum pressure of 2 GPa. The properties of obtained billets were compared to the properties of equidense specimens of the corresponding blends obtained under static pressing.

The characteristic temperatures of studied materials were determined by thermo-mechanical analysis (TMA) method with the TMI-1 plant at the load of 100 g. The effect of quantity and the dispersity of filler on physic-mechanical properties of composite materials (the density, the breaking strength, the breaking percent elongation, modulus of elasticity) were estimated after heat treatment at 380C according to GOSTs of 11262-80 and 93550-81.

As determined by the early investigations, the shock-wave activation of polymers creates the favorable conditions for interaction of active polymer centers with the filler surface [6]. The results of differential-thermal and thermal-gravity analysis of nickel-teflon and copper-teflon composite materials have demonstrated that explosion pressure, in comparison to traditional method s of obtaining the composite materials, causes the appearance of the stronger bonds between metal and teflon-4 [5,6].

The results of TMA of teflon-4 (fig.1) have demonstrated that for the specimens subjected to SWT (curve 2) the crystallite melting temperature decreases from 327 down to 280C and the maximum softening temperature decreases from 415-420 down to 355-360C in comparison to corresponding temperatures of initial polymer (curve 1). This is the evidence of physic-chemical modification of teflon-4 at explosion load, favoring the increase of rheological properties of material. The embedment of metallic filler causes the insignificant increase of melting temperature of teflon-4 crystalline formations and significant increase of softening temperature for the specimens obtained by explosion. Thus, for the materials subjected to SWT (curves 2, 4), the difference in temperatures between pure and the filled teflon is 70C, then for materials, statically pressed (curve 1, 3), this value is only 7C. That confirms the presence of the stronger adhesion interaction between metal and polymer for composite materials obtained by explosion pressure that, first of all, causes the limitation of free flow of CM in the region of the softening temperatures and is reflected in the change of material deformation ability.

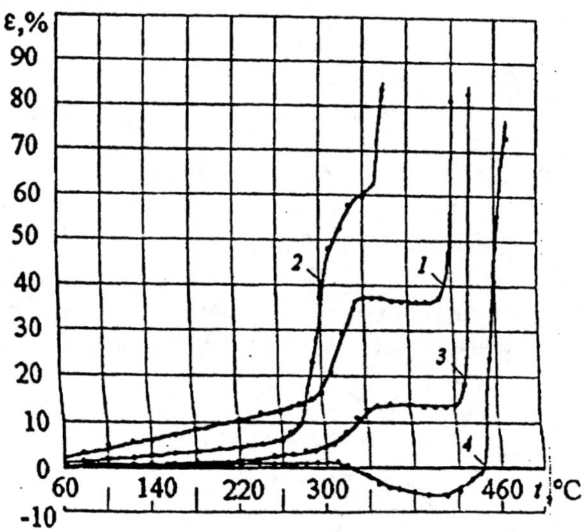

Figure 1. The thermomechanical curves of initial and filled teflon-4. 1 – the teflon-4, the static pressure; 2 – teflon-4, the explosion pressure; 3 – teflon-4+20%Cu, the static pressure; 4 – teflon-4+20%Cu, the explosion pressure.

The changes in the polymer structure, caused by high-speed deformation, reflected in the significant shifts of characteristic temperatures of physical processes, significantly increase the adhesion strength of composite materials on its basis that is confirmed by comparison of average values of mechanic properties of metal-teflon materials at the strengthening. As seen from the table, independently on the kind and quantity of metallic filler, the strength and plasticity of specimens after explosion pressure is higher than for statically pressed composite materials. For the latter, the increase of the amount of the filler up to 50% is followed with intensive decrease of composite strength. At SWT, due to the seizure bridges between metal particles the strength of composite material is constant. Furthermore, the formation of strong adhesion bonds between metal and polymer causes the increase of mechanical properties of composite materials.

Table: The Properties of Metal-Polymer Materials (Averages Values)

The amount of metal in composites with teflon-4	The stretching strength, MPa		Percent elongation, %	
	SP	EP	SP	EP
30%Cu	14	19	8	48
50%Cu	12	20	7	10
30%Fe	10	16	14	60
50%Fe	8	21	6	20
30%Ni	14	23	16	58
50%Ni	10	19	9	22

SP is the static pressure;
EP is the explosion pressure.

The study of effect of explosion treatment on the properties of teflon composites depending on the type (bronze, aluminum oxide), dispersity (14-250 μm), quantity (3-50%) of filler and the initial porosity of composite blend has demonstrated the following. As seen in fig.2, in the range of aluminum oxide concentration from 10 up to 40%, the strength of composites is practically constant or changes insignificantly, independently on size of filler particle (fig.2*a*). For whole range of concentrations the strength is higher for composites with the smaller particles. The plasticity of composites has the analogous dependence on dispersity of filler, sharply decreasing with increasing the amount of aluminum oxide. The modulus of elasticity at the stretching has the maximum for amount of aluminum oxide of 25-35%, the increase of dispersity of filler the maximum shifts in the region of smaller concentrations. The study of influence of bronze as the metallic filler on the mechanical characteristics of teflon composite materials at stretching has demonstrated the following. As seen in fig.3, the strength and modulus of elasticity of composite materials obtained by explosion pressure, starting from the amount of bronze of 20%, increases more intensively for the smaller filler. Such a character of dependence is explained by presence of the more developed surface in composites with fine-dispersed particles, favoring the increase of number of bonds. The more active interaction of metallic particles with polymer after explosion treatment allows additionally increase the strength of composites due to the increase of amount of bronze that is confirmed by character of change of plasticity. The present elongation of materials decreases with increasing the quantity of filler, but, in contrast to the composites filled with aluminum oxide, has higher values, approaching the pure teflon-4 for the small fill.

The differences in the dependence of physic-mechanical characteristics of composite materials on. the amount of aluminum oxide and bronze aluminum can be explained by different effect of explosion treatment and type of filler on the adhesion interaction of components. The shock-wave pressure of particles occurs at high rate in the regime near adiabatic. As the result, the temperature increases, which, due to a high plasticity and heat conductivity of bronze in comparison to aluminum oxide leads to the grip of metallic particles with carcass building that is reflected in the increase of strength and modulus of elasticity starting from 20% bronze and the decrease of the plasticity of composite with increasing the filler concentration. The heat energy for explosion pressure of teflon-4+alumin oxide composite, probably, is not enough for interaction of its more refractory particles that is confirmed by decrease of strength, the decrease of modulus of elasticity and more sharp decrease of plasticity than for bronze composite.

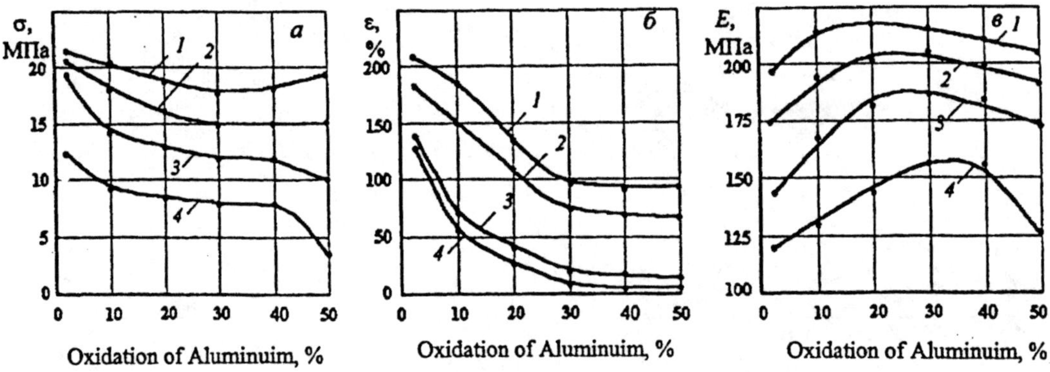

Figure 2. The dependence of strength (*a*), percent elongation (*b*), the modulus of elasticity (*c*) of teflon-4+aluminum oxide vs. dispersity and degree of filling. 1 – less 14 μm; 2 – less 50 μm; 3 – less 160 μm; 4 – less 250 μm.

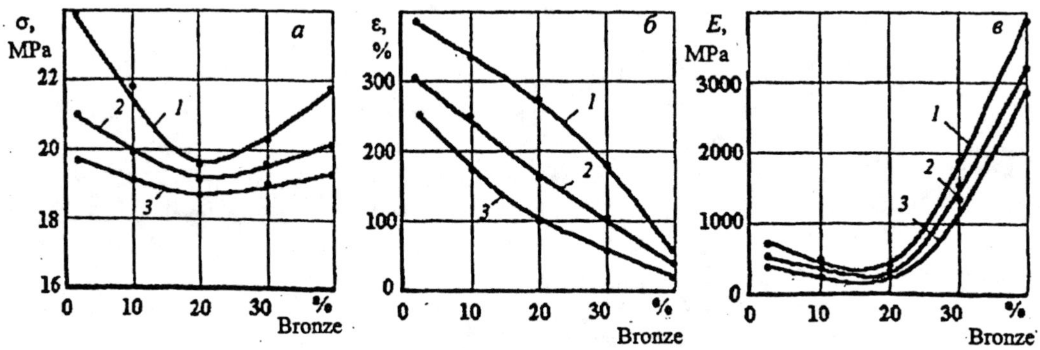

Figure 3. The dependence of strength (*a*), percent elongation (*b*) and modulus of elasticity (*c*) of teflon-4+bronze vs. dispersity and degree of filling. 1 – less 50 μm; 2 – 50-100 μm; 3 – 100-160 μm.

Beside, the presence of the stronger adhesion interaction of teflon-4 with the bronze than with aluminum oxide, can lead to the elements of polymer permolecular structure appearance in boundary layers, having a higher rigidity, and to the corresponding changes of properties of composite materials (the increase of strength and modulus of elasticity).

The effect of initial porosity on the physic-mechanical properties of CM has been studied for the teflon-4+20% bronze composite. The change of porosity of specimens was provided by preliminary prepressing of powder blend by the pressures of 0.5 up to 50 MPa that corresponded to the relative density at 20 up to 90%. The stretching experiments (fig.4) have demonstrated that independently on the filler dispersity, with decreasing the porosity the strength (a) and plasticity

(b) of composite materials gradually increase and reach the maximum at 50% of initial porosity. The densification above 60% leads to decreasing the σ and ε that confirms the influence of initial porosity on the thermal component of the explosion pressure process, the activation of polymer, the adhesion interaction on the boundary polymer-filler and, as the result, on the CM strength. The modulus of elasticity of composites of the different dispersity in a great extent depends on initial porosity and to the less extent – on the size of filler particles.

The used method of explosion treatment allows varying the amount of filler in the wide range. By changing the initial porosity of blend and the dispersity of filler, one can obtain the composites with different physic-mechanical properties that opens the prospective of using such a composite materials at producing the details and units of heightened reliability and duration.

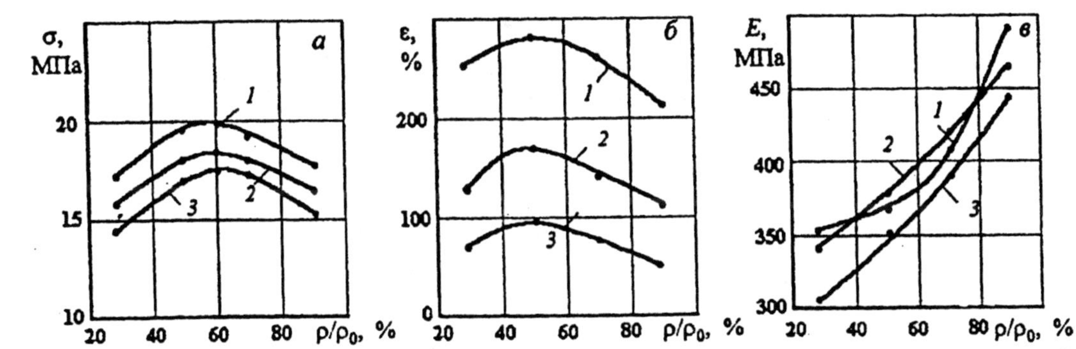

Figure 4. The effect of initial porosity and dispersity of metal-polymer blend teflon-4+20%bronze on the strength (*a*), percent elongation (*b*), modulus of elasticity (*c*) of CM: 1 – less 50; 2 – 50-100 μm; 3 – 100-160 μm.

The explosion treatment ambiguously effects on the separate mechanical characteristics of the obtained polymer composites and depends on the type, quantity and dispersity of filler. The strength and plasticity at stretching is significantly higher for composites with metallic than with mineral filler that confirms the more significant effect of heat rather pressure on the explosion activation.

CONCLUSIONS

1. It has been determined that explosion treatment changes the rheologial properties of teflon-4, increases its adhesion activity and physic-mechanical characteristics of polymer composites created on its basis by explosion pressure.
2. The maximum values of strength and plasticity of actual CM of tefln-4+bronze system is realized at 50% of initial porosity of mechanical blend.

REFERENCES

1. Korolev A.Ya. Himicheskoe modifitsirovanie poverhnosti tverdyh tel kak metod regulirovaniya ih adgezionnyh svoystv. Klei i tehnologiya skleivaniya. M.: Oborongiz, 1960, s. 35.
2. Mihalskiy A.I. Organofunktsionalnye apprety v napolnennyh polimernyh sistemah. Himiya i tehnologiya vysokomolekulyarnyh soedineniy. M.: VINITI, 1989, t. 19, s. 122-151.
3. Saadova A.N. Razrabotka protsessa modifikatsii PE karbokislosoderzhaschimi polimerami. Himiya i himicheskaya tehnologiya. M.: Izv. vyssh. uch. zav., 1976, № 8, s. 1258-1259.
4. 3itsans Ya.Ya., Kalnin M.M., Kronberga V.Zh. Ob izmenenii reologicheskih i strukturnyh harakteristik polietilena pri u-obluchenii v srede kisloroda. Modifikatsiya polimernyh materialov. Riga: RPI, 1978, vyp. 6, s. 126-138.
5. Adamenko N.A., Pavlov A.I., Trykov Yu.P., 3erschikov K.Yu. Strukturnye izmeneniya v polimernyh kom pozitah, obrabotannyh vzryvom. Fizika i tehnika vysokih davleniy. Kiev, 1995, s. 81-83.
6. Pavlov A.I., Adamenko N.A. Vozdeystvie udarnyh voln na metallopolimernye kompozitsii. V sb.trudov "Soveschanie po obrabotke materialov vzryvom" Novosibirsk, 1981,s. 174-176.
7. Adamenko N.A. Vliyanie vzryvnogo vozdeystviya na termicheskuyu obrabotku polimerov i ih kompozitsiy. Metallovedenie i prochnost materialov. Volgograd: VPI, 1990, s. 59-63.

ABOUT THE AUTHORS

Nina Aleksandrovna Adamenko – *the Volgograd State Technical University. A candidate of technical science, an associate professor. An expert in area of materials science of composite polymer materials.*

Yuri Trykov – *the Volgograd State Technical University. A Doctor in Technical science, a professor. An expert in area of materials science of composite polymer materials.*

Eduard Vasilievich Sedov – *the Volgograd State Technical University. An engineer, a postgraduate student. An expert in area of materials science of composite polymer materials.*

THE PECULIARITIES OF PHASE AND STRUCTURE TRANSFORMATIONS UNDER HIGH-TEMPERATURE HEATING OF Ni-Cr-Al-Y ALLOY CONDENSATE

V.A. Styazhkin, Yu.G. Veksler, A.M. Filippov and M.A. Filippov

The Ni-Cr-Al-Y system condensate materials were obtained by vacuum-plasma high-energy ions method. The structure and phase transformations in condensate materials under high temperature heating and interaction between coating and substrate have been investigated. The increase of protecting properties of constructional materials is the promising direction in creating the coatings.

The working performance of the lot details is determined not by the bulk but the surface properties. Therefore the investigation of combined systems consisting of the coating and the substrate is the actual task. The coating must provide corrosion and erosion resistance of surface and not effect on mechanical properties of material substrate. It is known [1] that the condensates obtained by means of deposition in the vacuum have the quasi-amorphous structure, confirmed by the strong line tailing on the difractograms. Under high-temperature heating of the coating after deposition, the main structural components are formed.

The structure and phase transformations in the Ni-20%Cr-11%Al-0.3%Y condensate on the ZhS5U steel substrate under long heating at temperature range of 1000-1100C have been investigated within frames of a given work.

THE MATERIALS AND INVESTIGATION TECHNIQUE

For obtaining the condensate coatings, the vacuum-plasma high-energy ion method has been used. The condensate structure is formed at depositing the ionized cathode material flux on the substrate specimen. The condensates have been obtained in MAP-1 industrial plant camera. The thickness of condensate coatings was 60-80 μm. The ZhS6U nickel alloy has been used as the substrate. The phase analysis has been carried out to difractometer DRON-3, the micro X-ray spectral analysis has been performed to the analyzer "CAMEBAX". The annealing of condensates was performed at 1000-1100C in vacuum furnace under pressure in the muffle under 10^{-5} Mp. The metallographic analysis has been performed to the microscope "EPITYP-II".

The coatings of nickel-chrome-aluminum system are widely used in the turbine construction as the hot canal detail protecting systems. After evaporation, the nickel-chrome-aluminum-yttrium condensate materials have the heterogeneous layer structure with the non-etching areas of width up to 15 μm (fig.1). The X-ray phase analysis exhibits the significant widening of the spectrum line that was classified as the X-ray amorphous structure. This is connected with the plasma-vacuum deposition rates. The basis peculiarities of a given kind of structure are the heterogeneity of chemical and phase composition crystalline lattice heterogeneity complex stressed state and, as sequence, the significant annealing temperature dependence.

The investigations of effect of the annealing temperature in vacuum at temperature range of 1000-1100C on the structure and phase composition of the coating have been performed. The results of X-ray analysis are presented in table 1. The chrome concentration in condensate surface layers decreases and, as the sequence, the γ- and γ'-phases are stabilized in this region. The annealing at 1000C causes the formation of face-centered cubic lattice of solid solution; the Cr β-phase emits from this solution under cooling (fig.2, the dark structure component).

Table: An Effect of Annealing Temperature on the Phase Composition of Ni-20%Cr-11%Al-0.3%Y Coating on the Zhs6u Steel Substrate

Annealing temperature, °C	Time, h	Phase composition of surface	Phase composition for depth of 30 μm
Without annealing	–	γ-phase, Ni_3Al, the traces of α-Cr	γ-phase, Ni_3Al, α-Cr, β-NiAl
1000C	4	α-Cr, β-NiAl, γ'-phase*	α-Cr, β-NiAl, γ'-phase, γ-phase
1100C	4	γ'-phase, β-NiAl	γ-phase, α-Cr, β-NiAl, $M_{23}C_6$-traces

*γ'-phase (light-etched component) is the γ-solid solution on the base of β-NiAl with α-Cr extractions.

Figure 1. A microstructure of Ni-20%Cr-11%Al-0.3%Y on the ZhS6U steel substrate after deposition, ×350.

The γ-γ' structure component is typical for the annealing temperature range of 1000-1050C. The active interaction in the boundary region occurs at annealing the condensate. For the Ni-20%Cr-11%Al-0.3%Y condensate on ZhS6U steel substrate, the structural composition changes with depth up to 30 μm (fig.3).

Figure 2. A microstructure of Ni-20%Cr-11%Al-0.3%Y on the ZhS6U steel substrate after annealing at 1000C, 4 h, ×350.

Figure 3 A microstructure of Ni-20%Cr-11%Al-0.3%Y on the ZhS6U steel substrate after annealing at 1100C, 4 h, ×350.

At exploring the coated details, the stability of properties of the protecting coating is determined by boundary factors. On the surface – it is the oxide film formation rate, on the steel boundary – it is the diffusion redistribution of elements [2]. In the region of steel boundary and the coating, the processes of diffusion interaction of two alloying systems develop. These systems are complex enough because of presence of several concentration gradients. In the region of steel bordering the condensate, the regions with the changed structure are formed (fig.3). The formation of these regions is connected, obviously, with chrome and aluminum diffusion into substrate under high-temperature endurance that causes changing the chemical composition of the steel. The enrichment of boundary regions of substrate with chrome causes shifting the dissolution range of γ'-phase into zone of lower temperatures, and the enrichment with aluminum causes increasing its volume portion [3]. The development of morphology occurs according to known mechanism of

"Ricks-Porter-Ecob" [4]. The interdendritic spaces are enriched with refractory alloy elements, creating the body-centered cubic extractions of α-W, Cr and carbides of M_6C type. Besides, the parts of γ-solid solution with the heightened concentration of refractory elements remain. The presence of such structural elements can lead to local emission of TPU-phases. As the result, under slow cooling followed after annealing, the structure consisting of γ'-phase with disperse extractions of γ- and α-Cr, α-W chrome-contained phases is formed. On other hand, the steel elements diffuse into condensate, the carbon, for an example, that is confirmed by presence of carbides in the middle part of coating after 25-hour endurance at 1100C.

Conclusions

1. The structure of Ni-20%Cr-11%Al-0.3%Y condensate on the ZhS6U steel substrate is formed under high-temperature annealing and is the fine-dispersed and heterogeneous one. The phase composition depends significantly on the temperature and the annealing time.
2. On the coating and substrate boundaries, the intensive diffusion processes connected with redistribution of alloying elements of the alloy and coating occur. This leads to penetration of alloy element into coating that worsens its heat resistance and reflects negatively on the strength.

The work has been performed under support of grant #2022 on fundamental problems of metallurgy of Ministry of general and professional education.

References

1. Styazhkin V.A., Kuznetsov V.P., Sokolov D.N., Solodar V.Ya. V sb. "Termicheskaya obrabotka i fizika metallov", Sverdlovsk, 1989, s. 135-139.
2. Movchan B.A-, Malashenko I.S. Zharostoykie pokrytiya, osazhdaemye v vakuume. Kiev: Naukova dumka, 1983, 228 s.
3. Sprengler C.J., Scheuer S.T., Barksdale D.C. Microstructural characterisation of servise exposed Co-Cr-Al-Y overlay coatings. Proc. 112th annu. meet., Atlanta, March 1983. Warrendale, 1983, p. 189-200.
4. Decker R.F. Strengthening mechanismus in nickel-base supperalloys. Steel strengthening mechanismus, Greenwich Corm Climax Molybdeniun, 1970, p. 147.

About the Authors

Victor Aleksandrovich Styazhkin *– the Ural State Technical University (the Ekaterinburg). A Candidate in Technical Science. An associate professor. An expert in area of constitutional steels and alloys, the protecting coatings.*

Yuri Genrikhovich Veksler *– the Ural State Technical University (the Ekaterinburg). A Doctor in Technical Science. A professor. An expert in area of refractory steels and alloys, the protecting coatings.*

***Aleksey Mikhaylovich Filippov** – the Ural State Technical University (the Ekaterinburg). A postgraduate student. An expert in area of constitutional steels and alloys, protecting coatings.*

***Mikhail Alekseevich Filippov** – the Ural State Technical University (the Ekaterinburg). A Doctor of Technical Science. A professor. An expert in area of constitutional materials.*

OBTAINING OF PLASMA POWDERS OF TITANIUM AND TUNGSTEN CARBIDES AND THEIR COMPOSITES

A.M. Lykov and A.L. Suris

The process of obtaining the titanium and tungsten carbides and their composites from chlorides in plasma-chemical reactor has been investigated. The thermodynamic analysis of these processes had defined the carbide existence region. The W_2C, WC_{1-x}, WC tungsten carbides, the titanium carbides and tungsten carbide solid solution in titanium carbide of different chemical and phase compositions and different dispersity, depending on process parameters (enthalpy, reagent ration), have been experimentally obtained.

The present work is devoted to investigating the synthesis of powders of superdispersed titanium and tungsten carbides and their composites, which are used in different areas of technics, for an example, in producing the solid solutions [1,2].

The carbides are synthesized in plasma-chemical reactor from the $TiCl_4$ and WCl_4 chlorides, those are feed in vaporous state via mixing camera as the flow system and then mixed with high-temperature hydrogen flow heated in plasma 20 kW power generator. After cylindrical plasma-chemical reactor the superdispersed carbide powder was catch by the fabric filter.

In order to determine the region of carbide phase existence, fractional conversion of initial substances into carbides, power-consuming and other theoretical indices of process of thermodynamic analysis of Ti-C-Cl-H and W-C-Cl-H systems has been carried out at different temperatures, pressures and ratio of elements in thermodynamically equilibrium system.

The tungsten single-carbide in the condensed state is present at temperature range of 1000-3500K. A more detailed thermodynamic analysis of these processes is presented in works [3,4].

As shown by experimental investigations of the process of obtaining the tungsten carbides, the synthesis of only tungsten WC single-carbide in condensed phase is practically impossible, since there is the semi-carbide W_2C and high-temperature cubic modification WC_{1-x} of different amount depending on the process regime parameters (the bulk enthalpy, the ratio of carbon-tungsten and other). In the region of low temperatures the tungsten can be as a powder.

The difficulty in obtaining the only carbide WC is obviously caused by that single-carbide has not region of homogeneity and exists only for metal and carbon stoichiometric ratio, and the carbonizing process passes through stage of obtaining the condensed tungsten or the W_2C, WC_{1-x} carbide phases, followed with carbon diffusion into the volume of the condensed particle which does not provide the stoichiometric ratio of the tungsten and carbon for whole particles. The quantity of bound carbon (which is in carbide crystalline lattice) does not practically exceed 4 mass.% (fig.1*a*). The enhancement of bulk enthalpy of energy carrier (hydrogen) leads to increasing the amount of combined carbon especially in region of $2\cdot10^4$-$3\cdot10^4$ kJ·kg^{-1}. Nevertheless, it is possible to obtain the powder with total amount of carbon (the combined in the carbide and the free), corresponding to carbon amount of 6.12 mass% in single-carbide which is turned to tungsten single-carbide without essential increasing the particle sizes, i.e. with maintaining the high level of powder dispersity.

The enhancement of bulk enthalpy of the energy-carrier is followed with intensifying the heat-mass exchange (diffusive) processes at formation of condensed phase that in this case is presented by the more carbon-saturated carbides.

As determined by X-ray phase analysis, for different $I=2\cdot10^4$, $3\cdot10^4$, $7\cdot10^4$ kJ·kg^{-1}, the basis tungsten-containing components are the W, W_2C and WC_{1-x}, respectively. The nature of carbidizer (methane or propane) does not effect practically on chemical and phase compositions of obtained powder. The increase of the carbon excess in initial reagents for seven times in comparison to stoichiometric one (W/C=1) does not increase its amount in the carbides (the combined carbon) and is equal to 4 mass, %.

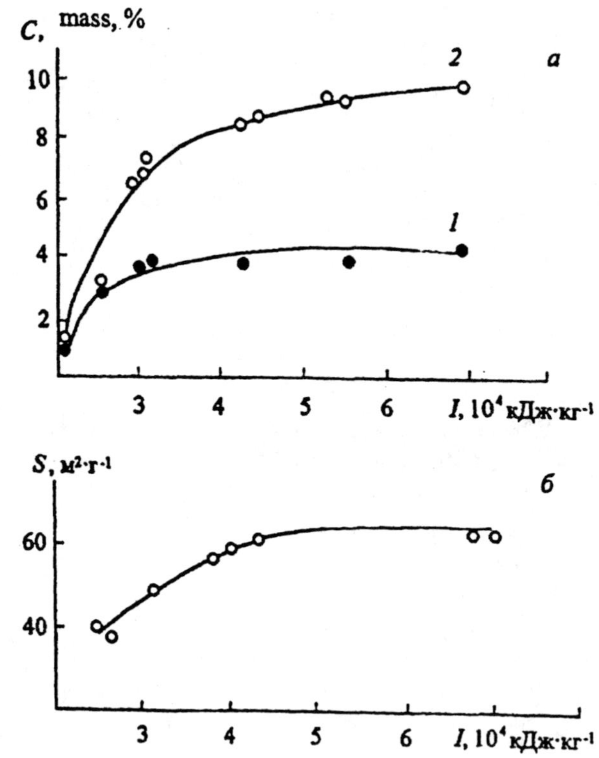

Figure 1. The influence of bulk enthalpy of energy carrier *I* on carbon amount in the condensed product C(*a*) and specific surface of powders S(*b*). 1 is the binder amount, 2 is the free carbon amount. The ration of the elements in the initial reagents is: C/W=6.4-8.2; H/C=47-79 (a), C/W=4.3-11.1; H/W=164-515 (b).

Thus synthesized powder consists of superdispersed particles, the specific surface increases from 40 up to 65 m^2·g^{-1} with energy carrier bulk enthalpy from 2.5 up to $7.2\cdot10^4$ kJ·kg^{-1} (fig.1*b*).

At synthesizing the titanium carbide in the plasma-chemical reactor, the superrdispersed powders of amount of combined carbon close to the stoichiometric one (18.8 mass.%) has been demonstrated. The amount of the combined carbide increases with the energy carrier bulk enthalpy, and for I>55 the powders consist of only titanium carbide. The lattice spacing of the obtained titanium carbide correlates good enough to the lattice spacing of the specimens synthesized by traditional methods practically for all region of homogeneity, presented in Storms' monograph

According to the microscopy analysis, the shape of the obtained titanium carbide powder particles is mainly cubic. By varying the bulk enthalpy, one can change the dispersity of obtained carbide practically for three times (fig. 2*a,b*). The specific surface of the titanium carbide powders vs. hydrogen enthalpy is presented in fig.3*b*. The average diameter of the particles of these powders, which was calculated as the volume-surface size on the basis of values of specific surface of the powder (sorption method) and the average diameter of the crystallite, determined by the method of frequency analysis of the X-ray diffraction line shape [6]. Thus, the sizes of particles, obtained by these methods correlate between each other that is the evidence that most particles of plasma-chemical titanium carbide is the single crystals.

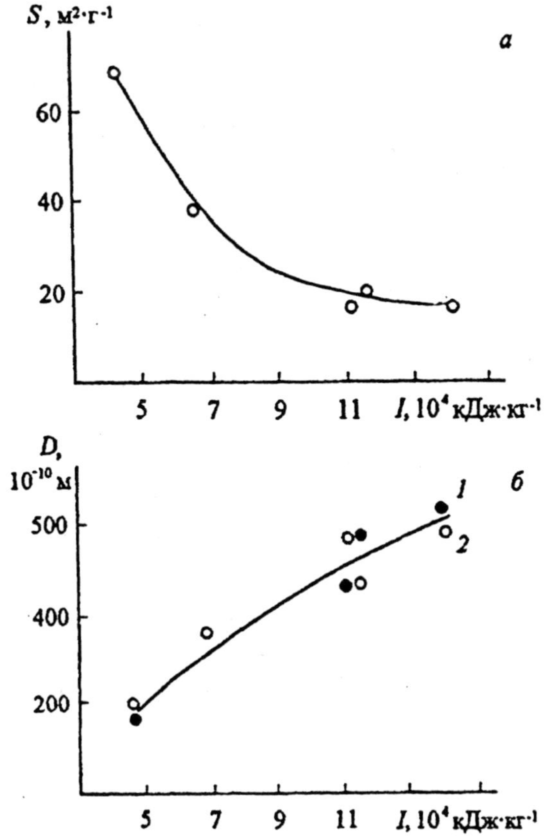

Figure 2. The influence of bulk enthalpy of the energy carrier on the specific surface S(a) and average size of particles, D(*b*). 1 is the sorption method, 2 is the method of frequency analysis of the X-ray diffraction lines.

The synthesis of the titanium and tungsten carbide composites was carried out for ratio of Ti/W near 1.5. Such a ratio of these elements is used for obtaining the solid alloys for the cutter on the basis of titanium and tungsten carbides. For atomic ratio in the initial reagents of $C/(Ti+W)_{init}<1.1$, the powder which does not contain the free carbon was synthesized.

The chemical analysis for the presence of titanium has demonstrated that ratio of Ti and W in the condensed product and in the initial reagent is practically identical, indicating that fractional conversion chlorides into metal in the condensed product is close to 1.

Atomic ratio of carbon to the number of tungsten and titanium atoms in condensed product, depending on the same ratio in initial reagents, is presented in fig.3. As seen in the presented dependence, the twelvefold excess of the carbon in the initial reagents practically does not effect on the amount of the combined carbon, being equal to 80% of the necessary for the stoichiometric composition, i.e. the face-centered cubic lattice of titanium carbide, on basis of which the solid solution of tungsten carbide in the titanium carbide is formed, has the deficiency of the carbon. The abundance of total carbon increases from 1.2 up to 1.6.

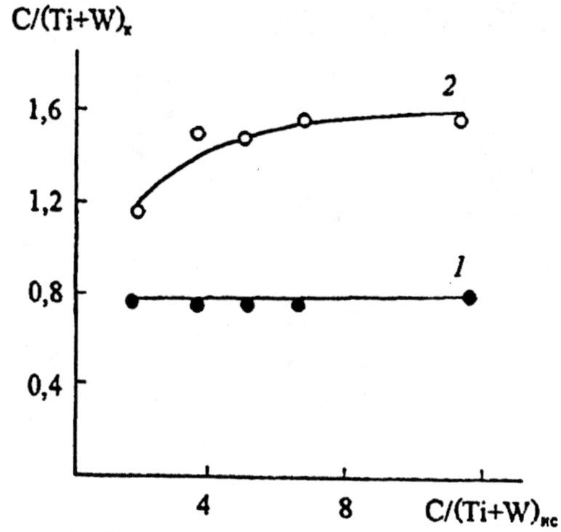

Figure 3. The influence of the relative excess of carbon in the initial reagents $C/(Ti+W)_{init}$ on its abundance in the condensed product $C/(Ti+W)_k$. 1 is the binder amount, 2 is total carbon amount. The bulk enthalpy I=56-60 $MJ \cdot kg^{-1}$, Ti/W=0.9-1.4.

The enhancement of the bulk enthalpy of energy carrier from $6 \cdot 10^4$ up to $10 \cdot 10^4$ $kJ \cdot kg^{-1}$ leads to increasing the carbon abundance in the powder up to 0.9, i.e. to the formation of the solid solution of tungsten carbide in titanium carbide, which is more close to the stoichiometrical one (fig.4). The abundance of the oxygen adsorbed by the powder from the air is $O/(Ti+W)<0.1$. Therefore, for obtaining the powders which do not contain the oxygen, it is necessary to create the hermetic systems for discharging it from reactor and the beyond revision (for example, the moulding and the sintering).

The amount of the chlorine in the product is $Cl/(Ti+W)=1.7 \cdot 10^{-2}$-$2.1 \cdot 10^{-2}$. In order to remove the chlorine from the powder it is necessary to anneal of the powder in vacuum.

At synthesizing the titanium and tungsten carbides, the solid solution of tungsten carbide in the titanium carbide is formed at combined carbon amount of $C_{com}/(Ti+W)=0.88$. The decrease of combined carbon leads to appearance of the reflexes of W second component in the difractograms. The intensity of W line increases with decreasing the abundance of the combine carbon down to 0.65.

The lattice spacing of plasma solid solution of tungsten carbide in titanium carbide corresponds well enough with data obtained for the powders of solid solution synthesized by traditional technology [7], and is equal to $4.314 \cdot 10^{-10}$ and $4.302 \cdot 10^{-10}$ m for the WC amount of 35 and 43 mole.% respectively.

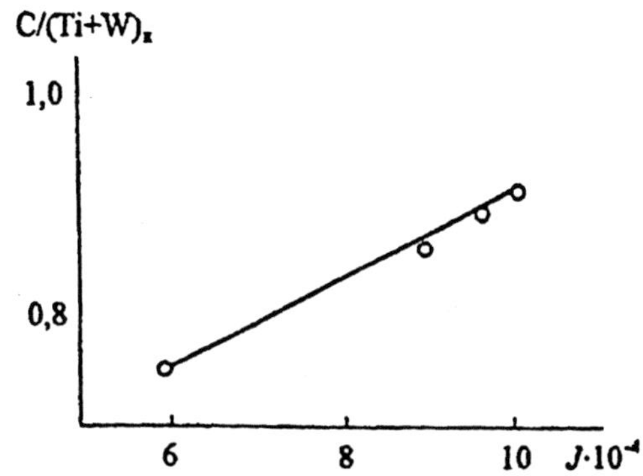

Figure 4. The influence of bulk enthalpy of the energy carrier on the abundance of the combined carbon in the C/(Ti+W) powder. The ration of the elements in the initial reagents is Ti/W=1.4-1.85; C/(Ti+W)=1.2-1.8.

The electron-microscopy analysis of the titanium and tungsten carbide composition powders has shown that the system is the aggregated superdispersed particles of $35 \cdot 10^{-9}$ m in the average volume-surface size (at specific surface of 18.1 m^2g^{-1}) that correlates to the size of particles, obtained from electron-microscopy analysis.

CONCLUSIONS

1. The possibility of obtaining the tungsten carbide in titanium carbide solid solution in the chlorides and hydrocarbon plasma-chemical reactor has been demonstrated.
2. By varying the operating parameters of plasma-chemical process, one can synthesize the superdispersed powders of titanium and tungsten carbides and their composites of different chemical, phase and dispersed compositions.

REFERENCES

1. Alekseev N.V., Samohin A.V., Kurkin E.I., Balihin I.L., Troitskaya E.V. Fizika i himiya obrab.mater., 1995, № 1, s. 31-39.
2. Tsvetkov Yu.V., Samohin A.V., Alekseev N.V., Agafonov K.N. V sb. Fizikohimiya ultradispersnyh sistem. Materialy IV Vserossiyskoy konf., M.: MIFI, 1998, s. 304.
3. Lykov A.M., Mamontov Yu.N., Panfilov S.A., Rezvyh V.F., Rykalin N.N, Suris A.L., Tsvetkov Yu.V., Shorin S.N. Fizika i himiya obrabotki materialov. 1979, № 2, s. 88-91.

4. Lykov A.M., Suris A.L., Shorin S.N. Termodinamicheskiy analiz protsessov polucheniya karbidov iz hloridov. II Vsesoyuznyy simpozium po plazme-himii. Tezisy dokladov. Зinatne, Riga, 1975, t. 2, s. 182-185.
5. Storms E. Tugoplavkie karbidy. Per. s angl. M.: Mir, 1977, s. 485.
6. Lykov A.M., Mamontov Yu.N., Suris A.L., Shorin S.N., Зefirova V.D., Plavnik A.L, Issledovanie dispersnosti plazmohimicheskogo karbida titana. Tezisy dokladov III Vsesoyuznogo simpoziuma po plazmohimii. M., 1979, s. 35-38.
7. Vitryanyuk V.K., Chaplygin F.N., Kostenetskaya G.D. Poroshkovaya metallurgiya, 1971, № 7, s. 46-52.

About the Authors

Aleksey Mykhaylovich Lykov *– the Small-Scale Research-Industrial Promotional Enterprise "MATTEX". A Candidate of Technical Science. The director. An expert in area of plasma treatment of the materials.*

Aleksandr Lasarevich Suris *– The Moscow State University of Engineering Ecology (MGUIE). A Doctor of Technical Science. A professor, a head of chair. An expert in area of plasma processes.*

INFLUENCE OF THE ALLOYING ON TUNGSTEN-COPPER ELECTRIC CONTACTS PROPERTIES

A.A. Notych, V.F. Krolokov and A.A. Zhilin

The effect of the controlled silicon alloying on tungsten-copper electric contact properties has been investigated. The density and the hardness of the KMK-B21 and BM70 (made in Bulgaria) electric contacts of different silicon amount after the pressing, the sintering and the correcting pressing are presented. The optimal percentage of the alloying dopants has been determined.

The properties of tungsten-copper pseudo-alloys, the electric contacts are significantly defined by the formation of the adhesion-firm contact of these metals. It is known that these metals do not chemically and diffusively interact each to other. Therefore the properties of the contact depends on wettability of the tungsten by the liquid copper that is the necessary condition of obtaining the qualitative material.

One of the main reasons of decreasing the adhesion strength of the phase boundaries is the segregation of oxygen atoms. The increase of the strength of phase adhesion bond is achieved by introducing the alloying elements which cause decreasing of the limiting wetting angle and increasing of the adhesion work.

The review of works on the composite materials of the pseudo-alloy type on the basis of W (Mo) and Ag(Cu) assigned for producing contacts of switching units is presented in [1]. The Ag, Co, Ni, Ti, Zr, Si, P3M, Y etc elements were introduced into composite materials as the alloying elements. The authors of [2] have established the character of change of properties in (W-Cu) (W-Cu+Ni) (W-Cu+Ag) (W-Cu+P3M) (W-Cu+Ti) (W-Cu+Zr) series. In this series the yield and the breaking point increases, and the wear resistance of materials increases as well [2]. The influence of controlled alloying with silicon of tungsten-copper pseudoalloys has been studied in works [3-5].

It is known that the oxides on the metal surfaces practically always worsen the wettability and increases the θ limiting angle. If all the oxides are moved from copper and tungsten off, then the strength of the phase boundary increases for five times as much even if θ is not equal to zero [6]. Therefore the fining with oxygen at producing the qualitative tungsten-copper electric contacts is the necessary condition for achieving their high mechanical properties. The fining with oxygen is achieved by using the reducing agent as well as by the controlled alloying of the copper with elements having a high affinity to the oxygen, such as silicon, yttrium, zirconium etc [1-5].

Despite of the existing information on the controlled alloying of pseudo-alloys with silicon [1,3-5], there is no information on the percentage of the alloying dopants. The silicon is not considered as the alloying element in work [2].

In present work, we have studied the alloying with the silicon on the density and hardness of electric contacts of the following compositions: 1) 26%Cu, 3%Ni, 70%W; 2)30%Cu, 3%Ni,

67%W. The percentage of components corresponds to the KMK-B21 and BM70 (made in Bulgaria) compositions of electric contacts [7]. The optimal percentage of the alloying addition has been determined as well.

The alloys were prepared in the following way: the cold pressing on the powder mixes, the liquid-phase sintering and the correcting pressing. The alloying dopants (silicon) was introduced into mixture in amount of 0;0.5;0.75;1;2;3%. The mixing of the tungsten PVT-u, copper PMS-1, nickel PNE-1 and silicon Kr1 powders was carried out in the ASP-3 planetary activator-mixer for 30 sec without using the balls. The pressing of the obtained mixture and the correcting pressing of the sintered contacts has been carried out to the DB2432A press with the nominal pressure of P_n=1.6 MN in the mold of 25.5×20.5 mm in the dimension of rectangular hole in the matrix. The pressure was varied within 950 Mp up to 1.6GPa. The correcting pressing was 2 GPa. Before the pressing and the correcting pressing the mold walls were lubricated with molybdenum disulfide. The density after the pressing and correcting pressing was calculated. The sintering of the obtained billets has been carried out in the vacuum furnace of the SNV1,3.1/16I model at temperature of 1250C for 1 h. In order to study the sintering kinetics, all the compositions were sintered in only hatchery. The density after sintering was calculated. The specimen hardness after sintering and correcting pressing was determined by the standard technique to the TR 5006-02 hardmeter by Rockwell technique, B scale.

Table: The Hardness of Electric KMK-B21 Contacts of Different Amount of Silicon after the Sintering and the Correcting Pressing

The composition of electric contacts	The hardness of electric contacts, HRB, after	
	the sintering	the correcting pressing
KMK-B21	20-25	40-50
KMK-B21+0.5%Si	66-68	86-88
KMK-B21+0.75%Si	73-75	90-95
KMK-B21+1%Si	73-78	85-90
KMK-B21+2%Si	77-78	–
KMK-B21+3%Si	–	–

The changes of the density after the pressing, the sintering and the correcting pressing are presented in fig.1-3.

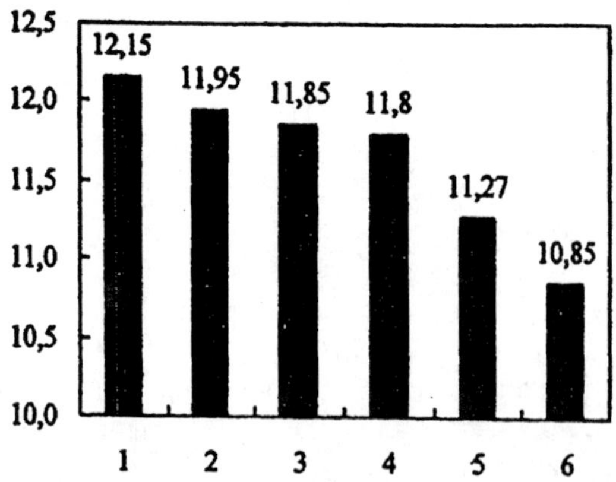

Figure 1. The density of contacts (g/cm^3) of different amount of silicon after pressing. 1 – KMK-B21; 2 – KMK-B21+0.5%Si; 3 – KMK-B21+0.75%Si; 4 – KMK-B21+1%Si; 5 – KMK-B21+2%Si; 6 – KMK-B21+3%Si.

The obtained results indicate that introduction of the alloying dopant decreases the density of electric contacts after pressing. With increasing the addition from 0.5 up to 3%Si the density of KMK-B21 contact decreases from 11.95 down to 10.85 g/cm^3 (fig.1). After sintering, the following picture is observed: the density of the silicon free electric contacts decreases, and the density of specimens with the alloying dopants (except 3%Si specimens) increases (fig.2). The specimens of the amount of alloying element of 3%Si are characterized with high level of cracking and, therefore, they were not subjected to the beyond investigations. The use of the correcting pressing significantly influences on the density and the hardness of the specimens with alloying dopant.

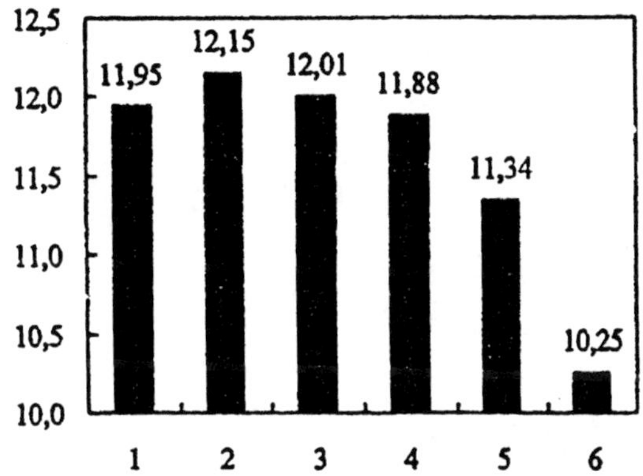

Figure 2. The density of contacts (g/cm^3) of different amount of silicon after sintering. 1 – KMK-B21; 2 – KMK-B21+0.5%Si; 3 – KMK-B21+0.75%Si; 4 – KMK-B21+1%Si; 5 – KMK-B21+2%Si; 6 – KMK-B21+3%Si.

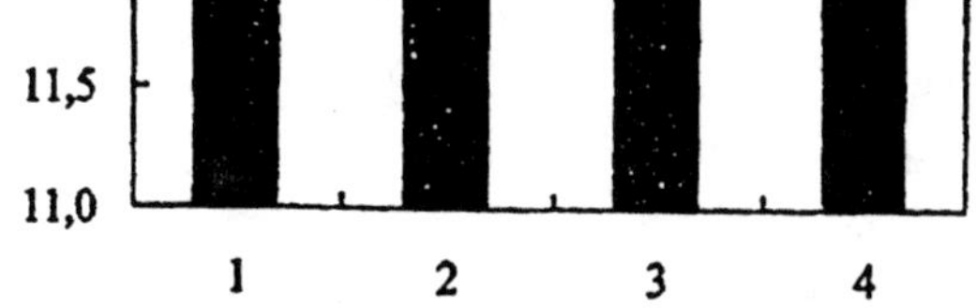

Figure 3. The density of contacts (g/cm^3) of different amount of silicon after the correcting pressing. 1 – KMK-B21; 2 – KMK-B21+0.5%Si; 3 – KMK-B21+0.75%Si; 4 – KMK-B21+1%Si.

For example, the density of KMK-B21 contacts, depending on the silicon amount, is within 12.7-13.15 g/cm^3, and the hardness is within 90-95 HRB.

According to the experimental results, the KMK-B21 and BM70 electric contacts of 0.75%Si have the optimal density and hardness.

The obtained results indicate on that the alloying with the silicon up to 1% significantly increases the hardness of tungsten-copper electric contacts. The results of a given work are in good agreement with the results of works [3-5] and allow assuming the following sequence of interaction of the tungsten with the copper-silicon melt.

After the melting, the silicon atoms near phase boundaries interact with tungsten and copper oxides, creating the silicates which are observed as the spherical; inclusions [3-5]. The interphase surface is refined and the strength of W-C contact increases. The silicon remains in the solid solution that is confirmed by increased hardness of the contact. The controlled alloying with the silicon also improves the plastic characteristics [3-5].

The results of a given work allowed optimizing the technological process of producing the electric contacts [8]. The electric contacts, alloyed with 0.75%Si, meet the requirements of GOST and are presently supplied to the Zaporozhye electric locomotive repair plant. The contacts have been experimentally verified and are mounted onto the control circuit of electric locomotives of VL80 series.

Conclusions

The controlled alloying with silicon significantly improves the wetting conditions and increases the adhesion work in the W-Cu system. The silicon refines the interphase surface and increases the hardness of tungsten-copper electric contacts. The introduction of the silicon more than 2% is assumed to be inexpedient because of high level of the cracking.

The optimal amount is 0.75%.

References

1. Minakova R.V., Grekova M. L., Kresanova A.P., Kryachko L.A. Poroshkovaya metallurgiya, 1995, № 7/8, s. 32-53.
2. Minakova R.V, Lesnik N.D., Kresanova A.P., Flis A.A., Homenko E.V. Poroshkovaya metallurgiya, 1996, № 7/8, s. 46-57.
3. Panichkina V.V., Pilipovskiy Yu.L., Radchenko P.Ya., Sirotyuk M.M., Ivaschenko Yu.N., Ponomarev S.S., Pereselentseva L.N. Poroshkovaya metallurgiya, 1986, № 4, s. 79-82.
4. Panichkina V.V., Sirotyuk M.M., Pilipovskiy Yu.L. Poroshkovaya metallurgiya, 1985, № 7, s. 43-46.
5. Sirotyuk M.M., Smirnov V.P., Panichkina V.V., Pilipovskiy Yu.L. Poroshkovaya metallurgiya, 1982, № 3, s. 101-105.
6. Nicholas N, Poole D. J. Mat. Sci, 1967, No 2, p. 269-274.
7. Spechennye materialy dlya elektrotehniki i elektroniki. Spravochnik pod red. prof. Gnesina. M.: Metallurgiya, 1981, 343 s.
8. Uluchshennye kontakty iz psevdosplava na osnove volframa. Dubina V.I., Krolikov V.F., Filippov D.A., Zhilin A.A.- Зaporozhe: Inform. listok Зaporozhskogo gos. tehyaich. universiteta, 1997.

ABOUT THE AUTHORS

Notych Aleksandr Anatolievich *– Zaporozhye State Technical University (ZSTU) (Zaporozhye, Ukraine). A Candidate in Technical Science. An associative professor. An expert in area of powder metallurgy.*

Krolokov Valeriy Fedorovich *– Zaporozhye State Technical University (ZSTU) (Zaporozhye, Ukraine). A research officer. An expert in area of powder metallurgy.*

Zhilin Andrey Anatolievich *– Zaporozhye State Technical University (ZSTU) (Zaporozhye, Ukraine). A postgraduate student. An expert in area of powder metallurgy.*

PHASE COMPOSITION OF HIGH-SPEED STEEL IRRADIATED BY HIGH-INTENSITY ION FLUXES

V.V. Uglov, D.P. Rusal'lskii, A.L. Kholmetskii,
V.V. Khodasevich, P.J. Wilbur and R. Wei

The R6M5 steel has been sequentially irradiated with high-intensive nitrogen and boron ion flux. As the result of implantation, the hardness has increased and the friction coefficient has decreased for two times. By investigating the element composition by Auger-electron spectroscopy and the phase composition of steel by Mossbauer conversion electron spectroscopy and by X-ray analysis, it has been determined that multi-layer structure with inclusion of the α"-$Fe_{16}N_2$, ε-Fe_3N nitrides and FeB, Fe_2B borides is formed in the steel.

INTRODUCTION

Presently, the method of high-intensity ion implantation (HII) into metals and alloys is violently developed [1-7]. This method allows creating the surface layers, saturated with metalloid ions which are bounded into nitrides or borides thus significantly increasing the performance characteristics of the treated material [1-7]. A given article is the continuation of the studies which have been published recently [7]. In this article we concentrate on the detailed study of phase composition of high-speed steel subjected to the sequent nitrogen and boron ions HII. For this investigation, the Mossbauer conversion electron spectroscopy has been used, thus the conversion electrons have been registered, allowing studying the layers of 0.1 μm of thickness, and the X-ray radiation of 10 μm of the scanning depth have been used as well. Thus, based on the obtained results, one can judge about phase distribution vs. depth of the implanted steel.

THE INVESTIGATION METHODS

The R6M5 steel is the subject of inquiry of a presented work. It is the high-speed steel of moderate heat resistance. The composition of the steel is as follows: 0.86%C; 6.0%W; 5.0%Mo; 4.1%Cr; 1.9%V; 0.5%Co (mass.%). The steel consists of martensite and carbides. The hardness of the steel is ~64 HRC. The steel specimens of the cylindrical shape of 15 mm in diameter and 2 mm in the height were subjected to thermal treatment to the standard technology for this steel class. After that they were grinded and polished to the only side by means of the diamond pastes of roughness of R_a=0.01 μm. Before irradiation the specimens were cleaned with ions.

The steel specimens were treated to the special plant [1,6]. The treatment procedure consists of the sequence ion implantation of nitrogen and boron ions at temperature of 500C. The irradiation regimes are presented in table 1. The specimen was heated by ionic beam. He temperature gradient within the specimen was about 10C. The nitrogen ionic beam is the blend $30\%N^{+}+70\%N_2^{+}$.

Table 1: The Implantation Regimes

#	Ion	Energy, keV	Current density, mA/cm^2	Irradiation time, min	Dose, ion/cm^2	Temperature, C
1	N	1	3.5	30	$7\cdot10^{19}$	500
2	B	20	0.53	10	$2\cdot10^{18}$	500
3	B+N	20+1	0.53+3.5	10+30	$2\cdot10^{18}+7\cdot10^{19}$	500
4	N+B	1+20	3.5+0.53	30+10	$7\cdot10^{19}+2\cdot10^{18}$	500

The concentration profiles of the alloyed and implanted dopants were determined with the help of the Auger-electron spectroscopy to the spectrometer PHI-660 (Perkin Elmer). The atomization has been performed by ion beam of energy of 3.5 keV.

The X-ray structure investigations (XRSI) were carried out by means of DRON-2 diffractometer in the geometry of the moving beam. The Co K_α characteristic irradiation of the wave-length of 0.17902 nm has been used. The hade of X-ray beam was 0.5 degree that corresponds to the specimen scanning depth of 0.15 μm.

The phase composition of initial and implanted steels was studied by means of conversion Messbauer electron spectroscopy (CMEM) using ^{57}Co/Cr source. Both the conversion electrons and the X-ray radiation have been registered using Ar-Xe proportional counter. In the first case, the surface layer of thickness of 0.1 μm, and 10 μm in second case has been scanned. The high sensitivity of CMES method gives the possibility of investigating the changes in the nearest surrounding of the atoms of Messbauer isotope (^{57}Fe in a given case) on the interatomic distances. It makes the CMES method more informative in studying the nearest order of amorphous phase in comparison to the methods of the transmission electron microscopy and X-ray structure analysis.

The microhardness measurements were carried out by Vikkers method to the PMT-3 micro-hardmeter at the load of 1N. The indenter penetration depth thus is about 2 μm.

The tribological experiments were carried out to the TAU-1 tribometer. The experiment scheme is: the lubricant free alternate motion and sliding friction. The contact is the "finger-flat". The indenter was produced from hard alloy VK8 (92%WC, 8%Co, in the mass.%) of the hardness of 87.5 HRC that exceeds the steel hardness. The bending radius of indenter is 0.5 mm. The load on the indenter is 2N.

EXPERIMENTAL RESULTS

The results of Auger-electron spectroscopy of steel specimen irradiated with nitrogen and boron ions are presented in fig. 1. It is seen that nitrogen is uniformly distributed. Its concentration is 9%, achieving the depth corresponding to 25 minutes evaporation. According to the profilometric measurements, it corresponds to 3.5 μm. The boron, in contrast to nitrogen, does not penetrate into the specimen depth. It remains on the surface layer of the thickness of 0.3 μm and its concentration achieves 80 atom.%. Thus, the N+B implantation has led to the formation of two-layer system, consisting of surface boron film and the deepened layer alloyed with nitrogen.

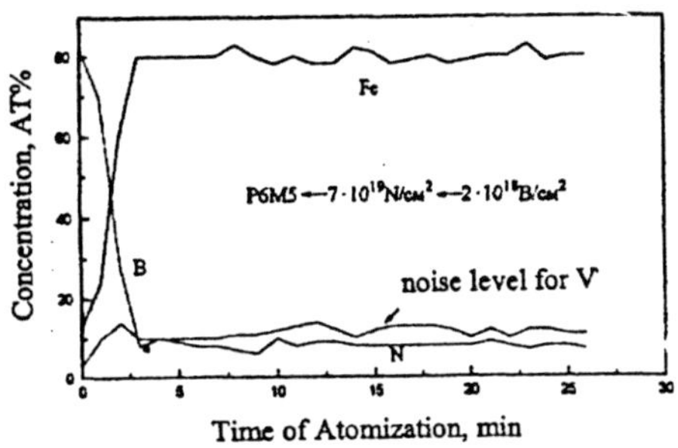

Figure 1. Concentration profiles of elements in R6M5 steel sequentially implanted with nitrogen and boron ions.

The parts of X-ray pictures of initial and implanted with nitrogen and boron ions of R6M5 steel are presented in fig.2. The X-ray pictures were obtained using the geometry of a moving beam at slide angle of 0.5 degrees. It corresponds to the scanning depth of 0.15 μm. On the X-ray picture of the irradiated steel, there is only diffraction line α-Fe (110) with the interplanar distance of d=0.2029 nm. After irradiation by N, N+B and B+N there is appeared one more peak with d=0.2092 nm, which is connected with ε-Fe_3N(101) nitride with d=0.209 nm. It should be noted that depth of the iron peak after irradiation decreases that is the evidence of the decreasing of the residual stresses in result of implantation.

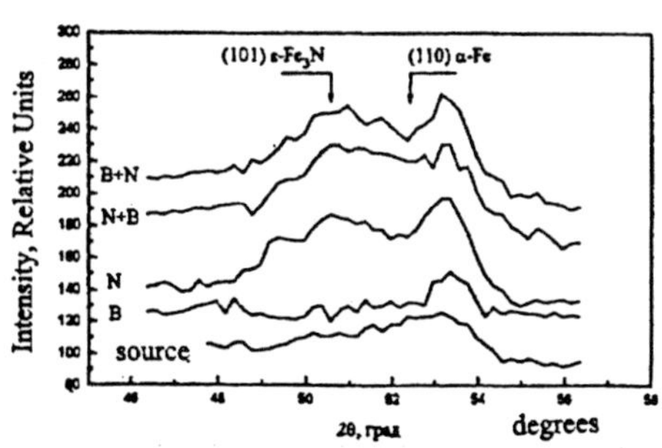

Figure 2. The regions of the X-ray pictures of the initial and the nitrogen and boron irradiated R6M5 steel. The angle of slide is 0.5 deg. The angle positions of the α-Fe(110) peaks of d=0.2029 nm and ε-Fe_3N(101) of d=0.209 nm are indicated in figure.

The CMES spectra of the initial R6M5 steel and the steel irradiated with high-intensive nitrogen and boron ion fluxes are presented in fig. 3*a-c*. The spectra presented in fig.3*b* and fig.3*c* are obtained at scanning the depths of 0.1 μm and 10 μm, respectively. The sub-spectra describing the presented phases are presented in the figures. The "interpretation" results with indicating the relative contribution of phases into CMES-spectra are presented in table 2 as well. The table 3 contains a super fine parameters of Messbauer spectra, described in fig. 3*a-c*. From figures and tables 2 and 3 it is seen that initial steel consists of martensite (94.7 at.%) and austenite (5.3 at%). The austenite concentration corresponds to that in the high-speed steel, passed the standard heat treatment [8]. In the result of irradiation with nitrogen and boron ions, the nitrides and borides are formed in the steel. The Messbauer parameters of the ε-Fe_3N and Fe_2B phases are near each to other [9,10] and are indistinguishable to the CMES. Therefore for multiple-Gaussian analysis they were described as the only phase. At scanning the steel up to the depth of 0.1 μm, the additional α"-$Fe_{16}N_2$ (13.7 at.%); ε-Fe_3N+Fe_2B (8.2 at.%) and FeB(4.1 at.%) phases were revealed. At studying the depth of 10 μm, the FeB phase was not revealed, the concentration of the rest phases in CMES-spectrum changes insignificantly. One should note on the behavior of martensite in the initial and irradiate steels. The sextet line depth, characterizing the martensite in the initial steel, is very large that indicates on the large number of non-equivalent positions occupied by iron atoms in the α-Fe(C) structure. As the result of the irradiation, the line depth of spectrum, describing the martensite, decreases, i.e. the martensite structure becomes more ordered.

The steel micro-hardness measurement results are indicated in the fig. 4. The measurements were carried out on both the steel surface and the cross thin-sections of the irradiated steel to the different distance from their edge. From these results one can judge about dependence of steel hardness vs. the depth. It is seen that for all the specimens the hardness is the largest on the surface. With increasing the depth, the micro-hardness of the irradiated specimens decreases, achieving the values of micro-hardness of the non-radiated steel. The specimens irradiated with N and N+B have the maximum value of the hardness for two times as much than for a non-irradiated steel.

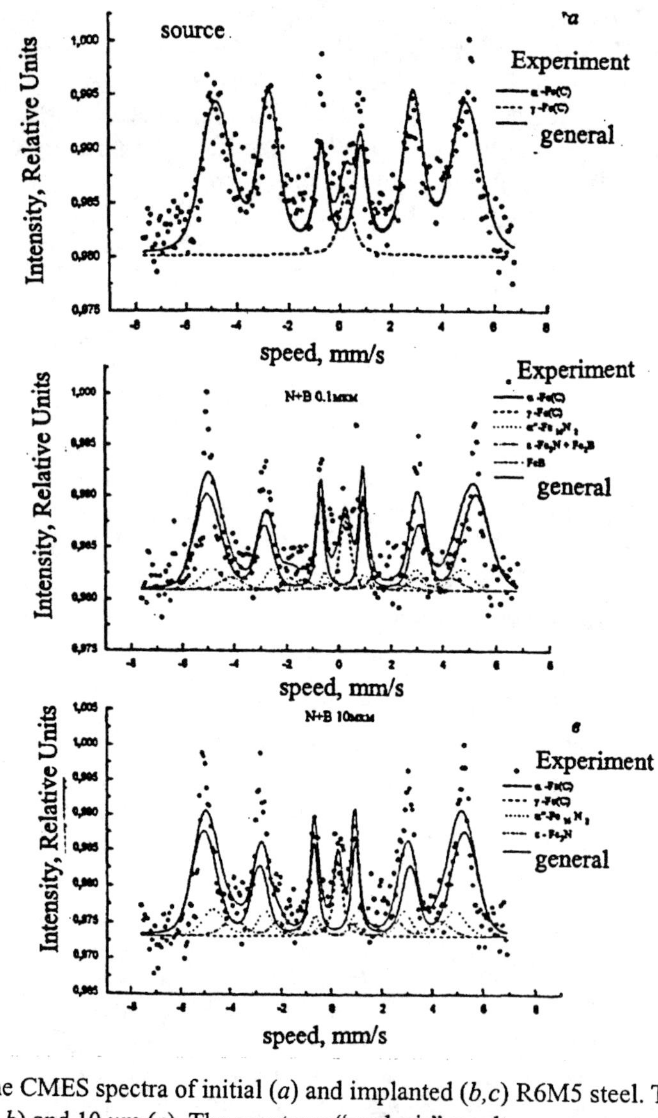

Figure 3. The CMES spectra of initial (*a*) and implanted (*b,c*) R6M5 steel. The scanning depth is 0.1 μm (*a,b*) and 10 μm (*c*). The spectrum "analysis" results are presented in the insets.

Table 2: The Phase Composition of R6M5 Steel Irradiated with High-Intensive Nitrogen and Boron Ion Fluxes ($7 \cdot 10^{19}$ N/Cm^2+$2 \cdot 10^{18}$ B/Cm^2)

The specimen	The phase composition, at.%				
	α-Fe(C)	γ-Fe(C)	α”-$Fe_{16}N_2$	ε-Fe_3N+Fe_2B	FeB
The initial	94.7	5.3	–	–	–
N+B 0.1 μm	67.5	6.4	13.7	8.2	4.1
N+B 10 μm	56.6	5.7	16.4	10.2	–

Table 3: The Mossbauer Parameters of Phases of R6M5 Steel, Obtained by High-Intensive Nitrogen and Boron Ion Fluxes ($7 \cdot 10^{19}$ N/Cm^2+$2 \cdot 10^{18}$ B/Cm^2)

The specimen	α-Fe(C)	γ-Fe(C)	α”-$Fe_{16}N_2$	ε-Fe_3N+Fe_2B	FeB
The initial	H_{eff}=30.24 Tl *IS*=0.109 mm/s *QS*=0.007 mm/s	*IS*=0.28 mm/s	–	–	–
N+B 0.1 μm	H_{eff}=31.84 Tl *IS*=0.100 mm/s *QS*=0.006 mm/s	*IS*=0.24 mm/s	H_{eff}=29.6 Tl *IS*=0.04 mm/s *QS*=-0.14 mm/s	H_{eff}=26.1 Tl *IS*=0.28 mm/s *QS*=-0.23 mm/s	H_{eff}=12.2 Tl *IS*=0.06 mm/s *QS*=0.26 mm/s
N+B 10 μm	H_{eff}=32.02 Tl *IS*=0.089 mm/s *QS*=-0.005 mm/s	*IS*=0.22 mm/s	H_{eff}=29.7 Tl *IS*=0.081 mm/s *QS*=-0.001 mm/s	H_{eff}=25.4 Tl *IS*=0.19 mm/s *QS*=-0.01 mm/s	–

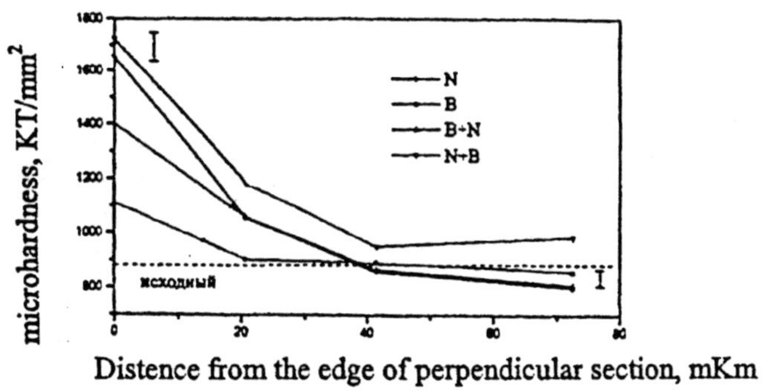

Figure 4. The microhardness of the cross thin section of R6M5 steel, implanted with nitrogen and boron ions on the different distance from its edge. The distance “0” corresponds to microhardness steel surface. The dashed line indicates the microhardness of initial steel.

The change of the friction coefficient of the non-irradiated and implanted steel in the dependence on the intender sliding path is presented in fig.5. It is seen that friction coefficient decreases in both the regime of the wear-in and the regime of the steady-state wear for the steel specimen, irradiated with N+B. The friction coefficient of this specimen under steady-state wear decreases for two times in comparison to the initial one.

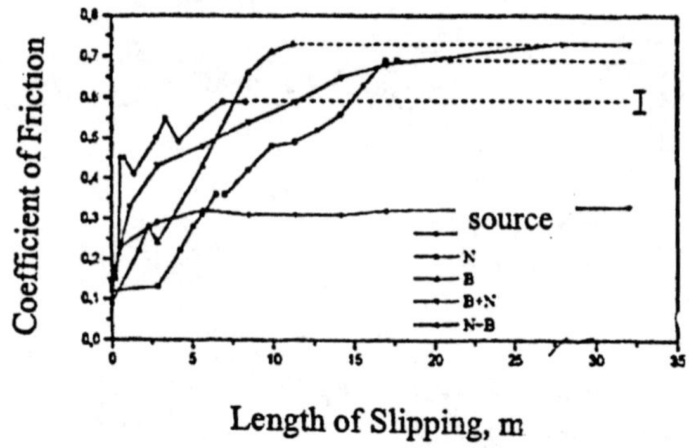

Figure 5. The friction coefficient of R6M5 steel implanted with nitrogen and boron depending on the friction path.

DISCUSSING THE EXPERIMENTAL RESULTS

From fig.4 and 5 it is seen that HII with nitrogen and boron ions worsens the mechanical properties f R6M5 steel. The largest increase of steel hardness and the decrease of friction coefficient for two times is observed for sequent $7\cdot10^{19}$ N/cm^2+$2\cdot10^{18}$ B/cm^2 irradiation The changes of phase and element composition of steel are responsible for such changes in mechanical properties. The results of element composition steel, irradiated with N+B, are presented in fig. 1. It is seen that nitrogen penetrates for 3.5 μm in depth (25 minutes of the evaporation). Recently [7] it has been determined that for the same nitrogen implantation of dose of $7\cdot10^{19}$ ion/cm^2 at temperature of 500C the nitrogen is able achieving the depth of 20 μm. The boron remains on the surface, forming the surface film. By means of XRSI it has been determined that ε-Fe_3N nitride of 0.15 μm of thickness is formed in the surface steel layer. Thus, in the result of HII with nitrogen and boron ions, the complex system, consisting of the deepened layer containing the ε-Fe_3N nitride, is formed in R6M5 steel with boron surface film. Under friction this film plays the role of solid lubricant, decreasing the friction coefficient, and the nitride layer increases the steel hardness (fig.4, 5). The CMES method with scanning the two layers of thickness of 0.1 μm and 10 μm has been used for a more detailed investigation of phase composition of the steel irradiated with N+B. The CMES investigation results can be compared to the results of XRSI measurements. The measurement results are presented in fig.3 and table 2,3. The CMES spectrum of the non-irradiated steel, presented in fig.3*a*, can be resolved into two sub-spectra: the singlet and sextet, characterizing the paramagnetic and magnetic phases, respectively. The singlet describes the γ-Fe(C) austenite of the concentration of 5 at.% that corresponds to the amount which must be contained in the high-speed steel after standard heat treatment [8]. Also steel must contain the martensite and about 20 mass.% of Fe_3W_3C carbides [8]. On the spectrum (fig.3a), the sextet describes superposition of martensite and carbide, since there is no data on this metastable carbide and it cannot be separated. It is seen that the lines of the sextet have the identical width that indicates the great number of non-equivalent states of iron atoms that is connected with strongly heterogeneous state of steel, containing a great number of defects and carbides and having the large residual stresses (fig.2). The average magnetic field, describing the magnetic sextet H_{eff}=30.24 Tl, corresponds to martensite, i.e. the Fe_3W_3C carbide, contributing into this sextet, must have a effective magnetic field of iron atoms, being near the martensite one. The confirmation of that is the CMES investigation results of the irradiated steel (fig.3,*b,c*). The recent investigations have shown [7] that due to the irradiation the great amount of carbide remains in the steel. These carbides must contribute to spectra; however the sextet, corresponding to the martensite, has the width of lines less than for non-irradiated steel. It indicates on, first of all, that H_{eff} of Fe_3W_3C carbide is near the martensite's one, and, secondly, in the result of irradiation, the steel structure is stabilized and steel surface layer is annealed. From fig.3 and table 2 it is seen that under irradiation, the α"-$Fe_{16}N_2$ (13.4%), ε-Fe_3N nitrides and FeB (4.1%), Fe_2B borides are formed in the steel surface layer of thickness of 0.1 μm. The ultra-fine parameters of ε-Fe_3N and Fe_2N phases are near those in [9,10]. Therefore it was not success to separate them by CMES method. Because of this, the Fe_2B and ε-Fe_3N phases were described in one-component CMES spectra, namely, by the sextet. The total amount of these phase is 8.2 atom.%. For the CMES investigations of steel layer of 10 μm, the FeB was not observed. That correlates well to the investigation of element composition (fig.1). The boron is in the surface layer of 0.3 μm, in which the pointed carbide is formed. At scanning the 10 μm layer, its contribution is negligibly small and, therefore, it does not contribute to the total spectrum in fig. 3*c*. In this connection, the Fe_2B boride must be absent as well, and only the ε-Fe_3N(10.2%) and α"-$Fe_{16}N_2$(16.4%) components, the α-Fe(C) and γ-Fe(C) martensite are present in fig.3*c*.

CONCLUSIONS

The high-intensive ion implantation of the R6M5 steel with nitrogen and boron allows forming two-layer system in steel, consisting of surface boron film with FeB, Fe_2B borides and the deepened layer, containing the α"-$Fe_{16}N_2$ and ε-Fe_3N nitrides. The high-intensive ionic implantation causes the arrangement of high-speed steel martensite structure. The two-layer system with inclusions of borides and nitrides in the surface steel layer increases its hardness and decreases the friction coefficient for two times.

The work has been reported on XXVIII International Conference on Physics of Interaction of Charged Particles with Crystals, Moscow, NII YaF MGU, 1998.

REFERENCES

1. Williamson D.L,, Ozturk 0, Wei R., Wilbur P.J. Surf. & Coat. Tech. 1994, v. 64, c. 15-23.
2. Wei R. Surf. & Coat. Tech. 1996, v. 83, c. 218-227.
3. Wei R., Vajo J.J., Matossian J., Wilbur P.J., Davis J.A,, Williamson D.L,, Collins G.A. Surf. & Coat. Tech., 1996, v. 83, c. 235-242.
4. Wilbur P.J., Davis J.A.,Wei R., Vajo J.J., Williamson D.L. Surf. & Coat. Tech., 1996, v. 83, e. 250-256.
5. Wei R., Shogrin V., Wilbur P.J., Ozturk 0., Williamson D.L. Journal of Tribology, 1994, v. 116, c. 870-876.
6. Wei R., Wilbur P.J., Ozturk 0., Williamson D.L. Nucl. Instr. and Meth., 1991, V. B59/60, c. 731-736.
7. Uglov V.V., Rusalskiy D.P., Hodasevich V.V., Kuleshov A.K., Rumyantseva I.N., Vey R., Vilbur P.Dzh. Fizika i himiya obrabotki materialov, 1997, № 6, s. 33-36.

8. Geller Yu.A. Instrumentalnye stali, M.: Metallurgiya, 1975, s. 572.
9. Jagielski J.,Kopcewicz M.,Thome L.J. Appl. Phys., 1993, v. 73, no. 10, p. 4820-4824.
10. Terwagne G., Piette M., Bertrand P., Bodart E. Science and Engineering, 1989, v. B2, p. 195-201

ABOUT THE AUTHORS

***Vladimir Vasilievich Uglov** – the Byelorussian State University (Minsk, Byelorussia). A candidate in Physics and Mathematics. An associative professor. An expert in area of ionic implantation and ion-plasma coatings.*

***Dmitrii Petrovich Rusal'lskii** – the Byelorussian State University (Minsk, Byelorussia). An engineer. An expert in area of multi-electron ionic implantation and the carbon films.*

***Aleksandr Leonidovich Kholmetskii** – the Byelorussian State University (Minsk, Byelorussia). A candidate in Physics and Mathematics. An Associative professor. An expert in area of structure-phase investigations.*

***Valerii Vsilievich Khodasevich** – the Byelorussian State University (Minsk, Byelorussia). A candidate in Physics and Mathematics. An associative professor. An expert in area of ionic implantation and ion-plasma coatings.*

***Paul J. Wilbur** – the Colorado University (Colorado, USA). An expert in area of high-intensive ionic implantation.*

***Rondhua Wei** – the Hodges Research Laboratories (California, USA). An expert in area of high-intensive ionic implantation.*

MURATAITE CERAMICS FOR ACTINIDES IMMOBILIZATION

B.S. Nikonov, B.I. Omel'yanenko, A.G. Ptashkin, S.V. Stefanovskii and S.V. Yudintsev

The new ceramics, based on murataite minerals, has been synthesized and the mineralogy-petrography analysis has been carried out. The new ceramics is good candidate as the matrix for immobilization of actinides and actinide-rare-earth fraction of highly active waste materials. The murataite composition corresponds to the nominal formula $A_4B_2C_7O_{22}$, where "A" position is occupied with Ca, Mn, TR and U cations; "B" is Mn, Ti, Zr and U; "C" is Ti, Al and Fe. There are two types of murataite structure – the fluorite with three- and fivefold fluorite motive reiteration of cell parameters of 14.8 and 24.6 Å, respectively.

INTRODUCTION

The actinides are the most dangerous components of the highly active waste materials (HAW). The half-life of most them is in the range from thousands to tens thousands years that defines the interval of their threat from a hundred thousands up to million years. For a safe immobilization of these elements the matrixes of a heightened resistance are required, being more durable than borosilicate or alumophosphate glasses, which are presently used for hardening the HAW. The search for the alternative materials has been provided since 50ies. At the end of 70-ies – the beginning of 80-ies, the crystalline matrixes consisting of chemical and radiation resistant phases, being able to include all the waste components, have been synthesized [1,2]. They consist mainly of the titanates (perofskite, hollandite, zirconolite, rutile), the oxides, the phosphates, the silicates and other crystalline silicates phase series. The Synroc ceramics is the most known among them [3]. Recently, there were created their species, oriented to the waste materials of the tailored fuel of civic (Synroc-C) and military (Synroc-D) reactors, the irradiated fuel (Synroc-F) and other. In American literature such composites are named as the "tailored nuclear waste ceramics". Since 1981 the Synroc-C is considered as the waste materials alternative to borosilicate glasses [4].

Lately, in many countries, including the Russia, the possibility of extracting the irradiated nuclear fuel Cs-Sr and actinide-rare-earth isotope groups for their separate hardening from liquid HAW in the matrixes of the optimal composition are investigated [5]. For immobilization of the actinide-consisted waste materials, including the plutonium, the ceramics based on zirconolite, pyro-clorines, perofskite, zircon etc are proposed [6]. Recently, at studying the ceramics of the waste materials of radiochemical plant "Mayak", the unusual phase of complex and variable composition has been revealed [7]. As the chemical and the structure properties, it was identified as the synthetic analogue of rare nature mineral – the murataite. The great attention was attracted to the uranium in the murataite, which is for 7-10 times as much than for zirconolite and perofskite. It allows considering the mineral as the possible matrix for fixation of the actinides. For verifying that assumption, the special experiments have been carried out, – those results are given in the present article.

The Preceding Investigation of the Murataite

The Natural Murataite

The murataite is the rare mineral, belonging to the complex oxides of titanium, niobium, rare-earth elements (REE), calcium and zinc. There are only three publications on the natural murataite, thus the two works are devoted to the same specimen. The first description has been performed for the specimen from Colorado, U.S.A. [8]. Its composition (table 1) corresponds to formula $(Na,Y,Er)_4Zn_2(Ti,Nb,Fe)_7O_{18}(F,OH)_4$. According to the roentgenometry, it has the cubic lattice of 14.86 Å of cell size. In work [9] the duplicate study of specimen from USA has been carried out, its spatial *F*4/3*m* group has been determined, being *a*=14.886(2) Å; the corrected formulas has been proposed: $(Y,Na)_6(Zn,Fe)_5\ Ti_{12}O_{29}(O,F)_{10}F_4(Z=4)$. This mineral has been revealed also in Baikal region, Russia [10]. In comparison to the specimen from Colorado, this mineral has a higher amount of calcium, zinc, titanium, being niobium and fluorine free, and has some different REE spectrum (table 1). The differences in the composition did not effect on the lattice size being determined as 14.87 Å. In general, even for two specimens, one can note the large variation of the composition of murataite and, consequently, the ability of isomorphous including of significant number of different elements, including REE. The structure of this mineral is suitable for the heavy rare earth metal embedment – the Dy-Lu and the yttrium.

Table 1: The Compositions of Natural and Synthetic Murataite (Mass.%)

Oxide	1[8]	2[9]	3[10]	4[11]	5[13]	6[14]
Na_2O	5.8	6.1	5.9	0.8	—	0.7
Al_2O_3	—	—	—	10.4	11.2	9.5
CaO	0.9	0.9	6.7	8.3	5.9	7.8
TiO_2	37.9	38.1	48.2	30.2	48.0	33.0
MnO	0.6	0.7	—	6.9	6.2	7.8
FeO	4.4	4.0	2.3	23.1	12.2	24.0
ZnO	12.5	11.3	16.2	—	—	—
Y_2O_3	12.1	12.5	20.5	—	—	—
$(La,Gd)_2O_3$	0.2	0.3	—	1.8	—	1.0
$(Tb\text{-}Lu)_2O_3$	9.5	10.5	0.7	—	—	—
ZrO_2	—	—	—	15.4	10.9	13.6
Nb_2O_5	10.0	10.1	—	—	—	—
UO_2	—	—	—	2.2	—	3.4
F	6.5	7.1	—	—	—	—
Total	100.4	101.6	100.5	99.1	94.4	100.8
Cell size, Å	14.86	14.886	14.87	14.576		

1-3 is the natural specimens; the synthetic specimens: 4 – is the calculation due to chemical formula [11], 5 – from [13], 6 is the calculation from atomic quantities [14]. The dash relates to the absent element.

Synthetic Murataite

The first artificial murataite was found in the ceramics developed for fixation of military HAW at Savanna River plant, USA [11-14]. They were prepared by sintering under 1150-1200C or by hot pressing at 1040-1200C and 14-69 MPa. The murataite was synthesized under different reducing conditions for oxygen pressure of 0.2 atm. (in the air) down to 10^{-14} atm. (CO_2-CO-graphite buffer). The murataite was revealed only in the specimens, obtained in the air – it is not stable under the reducing conditions. Its amount in the specimens is only 20-30%; the rest volume is filled with hollandite, nepheline, perofskite, zirconolite and spinel. The synthetic murataite differs significantly from the natural one in the composition (table 1). There is no zinc, niobium and fluorine; the natrium and rare-earth metal amount is the lowed; the calcium, manganese, iron and aluminum amount is higher; the zirconium is to the significant amount. The composition differences, probably, have defined the less cell size of the artificial phase (14.58Å) in comparison to natural murataite. Its composition is near the zirconolite one that in combination with small grain sizes complicates determination of their ratio.

The study of solubility of murataite consisted ceramics in the deionized water and in the solutions which are under equilibrium with the garnets, have demonstrated their high strength up to 400C [12,15]. The rate of uranium leaching from murataite matrix is 10^{-3}-10^{-4} g/(m^2·day). Only in the chloride solutions at T>300C the solubility of murataite increases, since the solution acidity increases due to hydrolysis of magnesium chloride.

STUDYING THE MELTED MURATAITE-CONTAINED CERAMICS

The murataite was found in the ceramics of the waste materials of "Mayak" plant [7]. It is placed among grains of early-created phases and grows up onto the zirconolite crystals (fig.1). For the composition near the zirconolite (table 2), the murataite is enriched with uranium and REE. The average amount of uranium and cerium is higher than in zirconolite for 7 and 4 time as much, respectively. The murataite, being only 5% of ceramics volume, accumulates the half of total uranium. One more peculiarity of a given phase is that the maximum amount of uranium is observed in the grain center and decreases to the boundaries (table 1). The study of solubility of murataite-contained ceramics in water at 350C and 50 MPa had demonstrated its high resistance to the leaching [7]. The uranium amount in the solutions after experiments was only 1-2 μg/l or <10^{-8} mole/kg of H_2O.

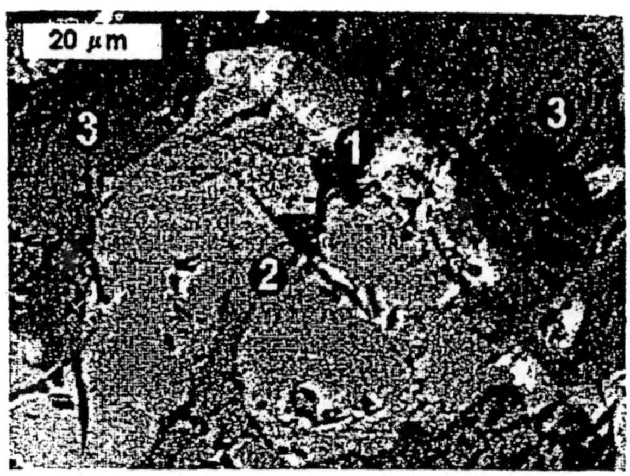

Figure 1. The SEM picture of the Synroc ceramics in the "Mayak" plant waste materials. The murataite grains white areas (1) grow up the zirconolite (2). The gray-dark is the hollandite (3).

The aim of this work was to study the ceramics with great amount of murataite, the study of structural peculiarities of mineral as well as the revealing of the main types of isomorphous replacements determining the variations of its chemical composition.

Table 2: The Average Compositions of Zirconolite (1), Perofksite (2), Murataite (3) in the Ceramics from "Mayak" Plant Waste Materials and the Area Compositions (4-7) of Single Murataite Grain (from the Center Forth the Boundaries)

Oxides	1	2	3	4	5	6	7
Al_2O_3	1.8	0.4	1.8	0.8	1.6	3.7	2.8
SiO_2	—	—	0.3	0.2	0.5	0.2	0.3
K_2O	—	—	—	—	0.3	0.7	—
CaO	12.6	33.5	15.6	15.6	16.5	13.0	15.9
TiO_2	45.9	57.4	59.8	54.8	56.5	62.9	65.9
Cr_2O_3	1.1	—	—	0.3	—	0.2	—
MnO	1.1	2.6	4.7	3.6	4.1	5.7	5.6
FeO	—	—	—	0.2	—	—	—
NiO	0.5	—	—	0.2	—	—	0.2
ZrO_2	34.4	0.2	5.6	5.7	6.0	3.8	4.3
BaO	—	1.7	0.6	—	3.4	3.7	—
Ce_2O_3	1.4	3.8	4.1	6.0	2.9	2.7	3.7
UO_2	1.0	—	7.0	11.9	7.0	2.0	0.8
The sum	98.8	99.6	99.5	99.3	98.8	98.6	99.5

THE TECHNIQUE

The synthesis has been carried out in alundum and platinum crucibles in the resistance furnace at 1400-1450C in the air. The duration for a given temperature is 10-100 minutes till mixture will be melted. After that, the furnace had been cooled down to 1000C for the rate of 10C/min in order to crystallize the melt. After that the furnace was turned out and it was cooled for 16 hours. There were performed six experimental series for the initial oxide mixtures of the following composition:

Ca-Ti-U-Mn-Al-Zr-Ce-O,

Ca-Ti-Mn-Al-Zr-Ce-O,

Ca-Ti-U-Mn-Zr-O,

Ca-Ti-U-Mn-O,

Ca-Ti-U-Mn-O,

Ca-Ti-U-Fe-O.

The first system corresponds to the murataite composition in the early-studied ceramics Synroc in the waste materials of "Mayak" plant [7]. Further, the element portion was excluded in order to determine those mixture components (except CaO and TiO_2), which are necessary for the murataite to be appeared. The specimens were studied to X-ray analysis and the JSM-53000 scanning electron microscope supplied with energy-dispersion spectrometer "Link ISIS". The analysis had been performed for the accelerate voltage of 25 keV and current of 1 nA, the zonde diameter was 1-3 μm, pulse set time is 50 seconds. The metals, the oxides and the phosphates were as the etalons. The threshold of element detectability is 0.2-0.4 mass.%, the analysis accuracy is

±10 rel.%. The murataite structure was investigated by means of electron micro-diffraction to the JEM-100c electron microscope.

DISCUSSING THE RESULTS

The chemical and phase composition of ceramics are given in table 3. Besides murataite, there were observed zirconolite, perofskite, rutile, hibonite, loverengite, pseudo-brukite as well as complex calcium, uranium and titanium oxide of pyro-chlorine structure. The phase composition depends on specimen composition: zirconolite is in zirconium high concentration specimens; rutile in titanium; pyro-chlorine in uranium. The murataite has been revealed only in the Mn-contained specimens, and its maximum amount (90%) is observed for the m5/2 and m2/5 ceramics. In the m3/1, m7/2, m5/4 and m2/4 specimens the murataite is under the subordinated amount, not exceeding 10-30%.

Table 3. A Calculated Chemical Compositions (Mass.%) of Specimens and Phase Composition of Ceramics from X-Ray Analysis and Electron Microscopy

Oxides	m3/1	m7/2	m5/2	m5/4	m2/4	m2/5	m3/3	m2/3
Al_2O_3	1.0	0.5	9.5	—	—	2.0	—	12.1
CaO	16.0	13.7	10.6	10.2	9.6	14.0	19.6	18.1
TiO_2	54.0	40.5	55.5	54.4	55.0	49.0	65.1	60.5
MnO	3.0	2.2	12.3	11.0	12.2	13.0	—	—
FeO	—	—	—	—	—	6.0	5.3	—
ZrO_2	7.0	9.6	8.2	3.5	—	7.0	—	—
Ce_2O_3	6.0	8.4	4.2	—	—	—	—	—
UO_2	13.0	22.1	—	20.9	23.2	9.0	10.0	9.3
The mineral phases	Py	Z	M	Py	Py	M	Pe	Py
(in descending order)	Pe	Pe	Z	M	R	L	Py	Pe
	R	M	H	R	M	R	R	R
	M	R	R		L	Ps	H	Ps

The dash is the element that did not introduce into mixture. H is hibonite, L is loverengite, M is the murataite, Ps is pseudobrukite, Py is Ca-U-Ti oxide with the structure of pyro-chlorine, R is rutil, Z is zirconolite.

By optical investigation it has been determined that murataite creates the crystals of the hundredth part of mm in size with square and isometric shape. The electron-microscopic study of murataite has demonstrated the changeability of its composition in the different specimens (table 4) and even at different regions of the same crystal. From he center to the grain boundaries, the amount of uranium, zirconium and cerium decreases; the amount of titanium, aluminum and manganese increases at practically unchanged calcium concentration. From the analytical definitions the aluminum has been revealed in the specimens, those were not doped with aluminum. It is obviously caused by that the ceramics synthesis was performed in the alundum crucibles being particularly dissolved in the melt that caused the contamination of the specimens with aluminum. The compositions of the synthetic murataite correspond to the formula $A_4B_2C_7O_{22}$, where A is Ca, Mn, TR, U; B is Mn, Ti, Zr, U; C is Ti, Al. This is near the formula of natural phase, which can be defined as $A_4B_2C_7O_{18}(F,OH)_4$, where A positions are occupied with Na, TR, Ca and Mn; B is Zn and Fe, and C isTi, Fe and Nb. The difference in the total anion charge is –44 for synthetic murataite, and –40 for the natural one is probably connected with that

B positions in the formula for synthetic phase are occupied with Ti^{4+}, Zr^{4+} and U^{4+}, and in the natural murataite – with bivalent zinc.

For the murataite of the ceramics from the waste materials of Savanna River plant, the other formula has been proposed: $A_2BC_7O_{16}$, A is Ca, Mn, TR; B is U, Zr, TR; C is Ti, Al, Fe. The unit cell size, being 14.58 Å for murataite from American ceramics, and 14.86-14.87 Å for natural murataite. Taking into consideration the above formulas, the main scheme of isomorphous exchanges in the structure of natural and synthetic murataite can be defined as follows:

in A position: Ca^{2+}=Mn^{2+} (synthetic),
$2Ca^{2+}$=Na^{+}+TR^{3+} (natural),
in B position: Ti^{4+}=Zr^{4+}=U^{4+} (synthetic),
Zn^{2+}=Fe^{2+} (natural),
in C position: $2Ti^{4+}$=$(Al,Fe)^{3+}$+Nb^{5+} (natural),
simultaneously, Ca^{2+}+Ti^{4+}=Na^{+}+Nb^{5+} (natural),
in the positions Ca^{2+}+Ti^{4+}=TR^{3+}+$(Al,Fe)^{3+}$ (natural),
A and C Ca^{2+}+$2Ti^{4+}$=Ce^{4+}+$2(Al,Fe)^{3+}$ (synthetic).

Table 4: The Composition (Mass%) of Murataite, Reduced to the 100%

Oxides	m3/1	m7/2	m5/2 (3ˣ)	m2/4 (5ˣ)	m2/5	m3/3	m2/3
Al_2O_3	1.4	0.2	24.3	7.4	6.6*	6.3*	4.9
CaO	15.5	13.3	7.8	10.8	6.9	7.0	10.9
TiO_2	54.0	41.5	47.5	52.5	48.5	48.3	48.2
MnO	4.5	2.5	10.7	10.1	20.0	22.3	9.8
FeO	—	—	—	—	—	—	4.7
ZrO_2	6.5	13.6	5.6	10.9	1.3	—	11.4
Ce_2O_3	11.0	19.6	4.1	8.3	—	—	—
UO_2	7.1	9.3	—	—	16.7	16.1	10.1

The dash relates to an absent element; * the appearance of Al_2O_3 in the m5/4 and m2/4 specimens is caused by dissolution of the crucible. 3ˣ and 5ˣ is the murataite with three- and fivefold fluorite cell.

The m5/2 specimen of the largest murataite amount (close 90%) has been studied to the transmission electron microscope. In the electron diffraction pictures, representing the (111)* murataite reciprocal lattice flat, the two nets of the different brightness intensity reflections are seen (fig.2). According to these data, there are two structurally different murataite species. For the murataite of first kind, between the most intensive and the central reflectance there are two less intensive reflexes. A given electron-diffraction corresponds to the cubic structure of fluorite type of 14.8Å cell size (the triple fluorite cell). The four less intensive reflexes are placed between strong reflectance in the pictures of murataite of second type. As follows, in this case, the cell parameter of murataite-like phase is equal to 25.6Å. The interplanar distances of the most intensive net reflexes are identical to those reflexes on the powder X-ray pictures (Table 5).

The composition of first structural species with triple fluorite cell is close to the murataite from American ceramics. In second murataite, the aluminum amount is smaller but zirconium and cerium amount is larger. The latter allows assuming that murataite with the similar structure has the larger heat capacity concerning the actinides. For confirming this assumption, the beyond investigations are required. In general, the complex, however, different kinds of isomorphism are realized in both the natural and the synthetic murataite. The latter is obviously to be connected with the difference of conditions of its formation in nature and experiment and, first of all, in the composition of mineral-formation medium. The natural murataites are the zinc minerals, and the manganese is constantly present in the synthetic phase composition. However, for some murataite analysis the amount of MnO decreases for less than 2 mass%, and the attempt of its synthesis from manganese free mixture have failed. In the Ca-Ti-U-Al-Fe-O system the cubic Ca, U and Ti oxide is synthesized. The X-ray characteristics of this oxide is: d=2.93, I=100; d=2.54, I=20; d=1.800, I=60; d=1.534, I=30; d=1.468, I=10; d=1.166, I=10, which corresponds to pyro-chlorine structure type.

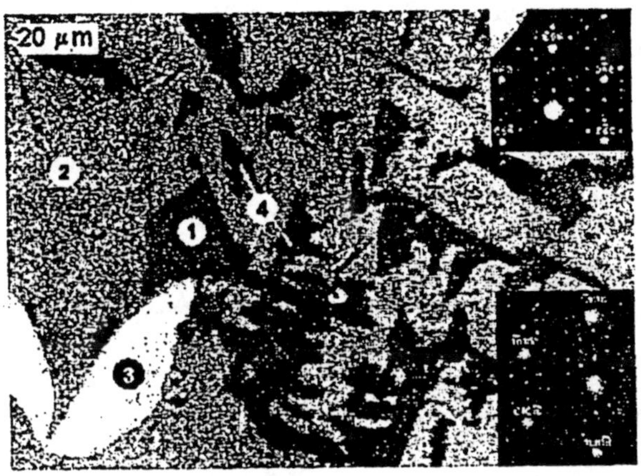

Figure 2. The SEM picture of the m5/2 ceramics with the murataite of two types (1 – with three- and fivefold fluorite cell). The white crystals to the down left part picture is the zirconolite (3), the small black isolations in the murataite-1 is the rutil (4). The incuts demonstrate a reciprocal lattice flat electronograms (111)* for both structural murataite species.

Table 5: The Diffractometer Data for Natural (1,2) and Synthetic (3,4) Murataite

1[8]		2[10]		3(m5/2 in table 3 and4)		The specimen #4 [11]	
d	*I*	*d*	*I*	*d*	*I*	*d*	*I*
8.51	30	—	—	—	—	8.44	3
4.29	5	—	—	4.25	5	4.205	2
3.028	10	3.01	10	—	—	—	—
2.858	100	2.86	100	2.847	100	2.803	100
2.468	60	2.48	40	2.464	18	2.430	21
2.141	20	2.13	10	2.134	4	2.101	6
2.080	20	2.08	10	—	—	2.04	6
1.746	80	1.75	60	1.740	40	1.715	43
1.489	80	1.496	50	1.483	22	1.466	33
1.432	50	1.430	20	1.420	5	1.466	33
1.138	50	1.139	20	1.128	4	1.148	9

The dash means that this line is not observed in the diffractogram.

The change of the murataite composition with respect to actinides and their imitators determines the following types of isomorphous exchanges: $Ca^{2+}+Ti^{4+}=TR^{3+}+(Al,Fe)^{3+}$, $Ti^{4+}=Zr^{4+}=U^{4+}$, $2Ca^{2+}=Na^{+}+TR^{3+}$. The same replacements are typical for other actinide-contained phases of radioactive wastes – the zirconolite and perofskite [3,6,16,17]. Our data exhibit the ability of murataite to include up to 30 mass% of isotopes of actinide-rare-earth fraction of HAW. That, obviously, is caused by the presence of several structural positions ensuring the embedment of variable valence from +1 (natrium) up to +5 (niobium) and ion size from 0.5-0.6Å(Al^{3+}, Nb^{5+}) up to 1.0-1.2Å(Na^{+}, Tr^{+}, U^{4+}). The values of the corresponding parameters for actinide ions are in these ranges. As mentioned above, the same is valid for the zirconolite structure. The comparison of murataite and zirconolite properties in the light of prospective of their usage for the actinide immobilization are presented below.

Both the zirconolite and murataite minerals are the complex oxides material based on titanium, calcium, rare earths etc and have a good structure derivative to the fluorite lattice type. The phase structure contains several positions providing with isomorphous embedment of the varied radius and charge cations. The murataite and zirconolite have a great isomorphous capacity with respect to actinides and their imitators. The total amount of uranium and REE in these phases reaches 30 mass.%. They are chemically resistant to the hydrothermal solutions treatment. At the same time, there is some difference between them. The zirconium is the basis component in the zirconolite. At least 70% of total number of places in "B" position of ABC_2O_7 zirconolite must be occupied with zirconium [3]. The Zr_2O amount in the synthetic and natural zirconolites, as a rule, is not less of 25 mass.%. The zirconium is not necessary component for formation of the murataite, and its natural species are zirconium free. Besides of calcium and titanium, the necessary element for murataite formation is obviously zinc (in nature) or manganese (in experiments), thus amount of the latter can be the little percents. One more difference is the peculiarities in the distribution of uranium in minerals. The uranium is uniformly distributed for whole zirconolite crystal; in the murataite the uranium is concentrated in the center grain parts. It creates one more barrier, resisting to the element leaching. The uranium distribution between co-existing murataite and zirconolites shifted in favor of the murataite. The uranium amount in the parts of murataite phase formed simultaneously with zirconolite is higher than in zirconolite. In nature, the zirconolite was found in quartz-contained rocks, as the zirconolite was not found in such medium. It is probably to be caused by its instability under silica high intensity conditions, when the zirconium is a stable phase. Therefore the murataite ceramics can appear to be more solid than the materials on the basis of zirconolite at building the burial, in the quartz-contained rocks, for example, in granites or tuffs. At the same time, the lack information lack on the murataite does not allow carrying out the complete comparison of murataite and zircnolite materials properties. For that, it is necessary to study the embedment of real actinides – Np, Pu, Am, Cm into murataite structure, as well as to investigate its radiation and chemical stability. After obtaining such an information, one can more precisely estimate the serviceability of murataite for actinide-rare-earth HAW fractions immobilization.

CONCLUSION

The ceramics, mainly consisting of murataite – complex titanium, calcium, manganese, aluminum, uranium and rare-earth metal oxides of fluorite structure has been synthesized. The murataite composition corresponds to nominal formula $A_4B_2C_7O_{22}$, where A position is occupied with Ca, Mn, TR and U cations; B is Mn, Ti, Zr and U; C is Ti, Al and Fe. The existence of structural murataite species of thee- and fivefold repetition of the basis fluorite motive of cell parameter of 14.8 and 24.6Å, respectively, have been revealed. The different types of the exchanges, including the participation of actinides and REE, are typical for murataite lattice. The zirconolite has the similar structural replacements, however the distribution of uranium and REE between these phases is shifted in favor of murataite. The uranium concentration in the center murataite grains is significantly smaller than on the boundaries. It provides the additional barrier resisting to the leaching of radionuclids from murataite ceramics. The obtained results indicate on the necessity of beyond study of murataite ceramics as potential matrix for fixation of actinide-rare-earth fraction of HAW.

Authors are thanking to M.I.Lapin, A.I.Gorshkova and A.V.Sivtsova (IGEM RAS) for the help in carrying out the electron microscope investigations.

REFERENCES

1. Ringwood A.E., Kesson S.E., Ware N.G., et al. Nature, 1979, no. 278, p. 219-223.
2. Harker A.B, Morgan P.E.D., Clarke D.R., Flintoff J.F. Sci. Basis for Nucl. Waste Manag., 1986, v. 6, p. 567-574.
3. Ringwood A.E., Kesson S.E., Reeve K.D., Levins D.M., Ramm E.J., SYNROC, Radioactive Waste Forms for the Future, eds. W. Lutze and R.C. Ewing, Elsevier Science Publishers B.V, 1988, p. 233-334.
4. Hench L.L, Dark D.E., Campbell J. Nucl. Chem.Waste Manag., 1984, v. 5, p. 149-173.
5. Glagolenko Yu.V., Dzekun E.G., Drozhko E.G., Medvedev G.M., Rovnyy S.I., Suslov A.P. Vopr. Radian. 1996, no. 2, s. 3-10.
6. Ewing R.C., Weber W.J., Clinard F.W. Progress in Nucl. Energy, 1995, v. 29, p. 63-127.
7. Sobolev I.A., Stefanovsky S.V., Loudintsev S.V. et al. Mat. Res. Soc. Symp. Proc., 1997, v. 465, p. 363-370.
8. Adams J.W, Botinelly T, Sharp W.N., Robinson K. Am. Miner., 1974, v. 59, nos. 1/2, p. 172-176.
9. Ercit T.S., Hawthorne F.C. Canad.Miner., 1995, v. 33, p. 1223-1229.
10. Portnov A.M., DubakinaL.S., Krivokoneva G.K. DAN SSSR, 1981, t. 261, s. 741-744.
11. Morgan P.E.D., Ryerson F.J. J. Mater. Sci. Lett, 1982, v. 1, p. 351-352.
12. Morgan P.E.D., Harker A.B., Flintoff J.F., et al. Proc. 2-nd Int. Symp. on Ceramics in Nucl. Waste Manag. "Advances in Ceramics". Chicago, 1983, v. 8, p. 234-246.
13. Ryerson F.J. J. Amer. Ceram. Soc., 1983, v. 66, p. 629-636.
14. Ryerson F.J. J. Amer. Ceram. Soc., 1984, v. 67, p. 75-82.
15. Freebom W.P., White W.B. Proc. 2nd Int.Symp. on Ceramics in Nucl.Waste Manag. "Advances in Ceramics". Chicago, 1983, v. 8, p. 368-376.
16. Vance E.R., Agrawal D.K. Nucl. and Chem. Waste Manag. 1982, v. 3, p. 229-234.
17. Lumpkin G.R., Smith K.L., Blackford M.G. J. Nucl. Mater. 1995, v. 224, p. 31-42.

ABOUT THE AUTHORS

Boris Sergeyevich Nikonov *– the IGEM of RAS. A research officer. An expert in area of mineralogy and petrography.*

Boris Ivanovich Omel'yanenko *– the IGEM RAS. A Candidate of Geology and Mineralogy. A laboratory chief. An expert in area of uranium geochemistry, the geology of uranium fields, mineralogy and petrography.*

Aleksandr Gennad'evich Ptashkin *– the Russian Chemistry-Technological University after D.I.Mendeleyev. A postgraduate student. An expert in area of development and synthesis of new forms for binding the radioactive waste materials.*

Sergey Vladimirovich Stefanovskii *– the Moscow Scientific Industrial Plant RADON. A professor, a doctor in Chemistry. A deputy director of Research Center of Genecology. An expert in area of physics and chemistry of materials binding the radioactive waste.*

Sergey Vladimirovich Yudintsev *– IGEM of RAS. A Candidate of Geology and Petrology. A Research Officer. An expert in area of mineralogy and petrography, including the ceramic forms of radioactive waste.*

INDEX

D

E

F

G

R

S

T

U